Jan-Philip Jarvis

A Contribution to Active Infrared Laser Spectroscopy for Remote Substance Detection

Schriftenreihe Automatische Sichtprüfung und Bildverarbeitung
Band 10

Herausgeber: Prof. Dr.-Ing. Jürgen Beyerer

Lehrstuhl für Interaktive Echtzeitsysteme
am Karlsruher Institut für Technologie

Fraunhofer-Institut für Optronik, Systemtechnik
und Bildauswertung IOSB

A Contribution to Active Infrared Laser Spectroscopy for Remote Substance Detection

by
Jan-Philip Jarvis

Scientific Publishing

Dissertation, Karlsruher Institut für Technologie
KIT-Fakultät für Informatik

Tag der mündlichen Prüfung: 27. Juli 2017
Referenten: Prof. Dr.-Ing. habil. Jürgen Beyerer
Prof. Dr. rer. nat. habil. Oliver Ambacher

Impressum

KIT Scientific Publishing

Karlsruher Institut für Technologie (KIT)
KIT Scientific Publishing
Straße am Forum 2
D-76131 Karlsruhe

KIT Scientific Publishing is a registered trademark
of Karlsruhe Institute of Technology.
Reprint using the book cover is not allowed.

www.ksp.kit.edu

Print on Demand 2017 – Gedruckt auf FSC-zertifiziertem Papier

ISSN 1866-5934
ISBN 978-3-7315-0725-3
DOI 10.5445/KSP/1000074148

A Contribution to Active Laser Spectroscopy for Remote Substance Detection

Zur Erlangung des akademischen Grades eines

Doktor der Ingenieurwissenschaften

der Fakultät für Informatik
des Karlsruher Instituts für Technologie (KIT)

genehmigte

Dissertation

von

Jan Jarvis
aus
Freiburg im Breisgau

Tag der mündlichen Prüfung: 27. Juli 2017
Referent: Prof. Dr.-Ing. habil. Jürgen Beyerer
Korreferent: Prof. Dr. rer. nat. habil. Oliver Ambacher

Zusammenfassung

Die abstandsfähige Detektion von Gefahrstoffspuren war in den letzten Jahren Gegenstand intensiver Forschung und stellt dennoch nach wie vor ein ambitioniertes Ziel dar. Zur Lösung des Problems wurden verschiedenste Methoden vorgeschlagen, unter denen derzeit die erfolgversprechendsten zweifellos die laserbasierten Techniken sind.

In dieser Arbeit wird ein augensicheres und abstandsfähiges Infrarotspektroskopiesystem für die berührungslose Detektion von Gefahrstoffen vorgestellt. Das vorgeschlagene Messprinzip basiert dabei auf wellenlängenselektiver Beleuchtung mittels breitbandig abstimmbarer Quantenkaskadenlaser (QCL) im langwelligen Infrarotbereich zwischen 7.5 µm und 10 µm. Dieser Spektralbereich ist Teil des molekularen Fingerabdruckbereichs für viele chemische Verbindungen, so auch für Explosivstoffe wie PETN, RDX, TNT und AN.

In der bildgebenden Variante des hyperspektralen Messsystems wird das diffus rückgestreute Licht mittels einer hochempfindlichen Quecksilber-Cadmium-Tellurid (MCT) Kamera eingesammelt. Durch synchrones Abstimmen der QCL Emissionswellenlänge wird ein Hyperspektralbild der beleuchteten Szene erfasst, bei der jeder Pixelvektor dem Rückstreuspektrum eines bestimmten Ortes entspricht. Dieser Pixelvektor wird als lineare Superposition der beteiligten Materialspektren oder *Endmember* angenommen, bei der die relativen Gewichtungen nicht bekannt sind. Die Menge aller Endmember im Hyperspektralbild kann im Allgemeinen den reinen Materialien in der Szene zugeordnet werden. Das Ziel der Materialdetektion in Hyperspektralbildern besteht darin, bekannte Substanzspek-

tren in den erfassten Hyperspektraldaten zu detektieren und zu identifizieren. Viele bekannte Methoden zur Lösung dieses Problems basieren auf dem Linearen Mixtur Modell (LMM), dass die lineare Superposition der beitragenden Endmember voraussetzt.
In dieser Arbeit werden Details des Hyperspektralbildsensors diskutiert, die sowohl Optimierung der Beleuchtung als auch Erfassung der Messdaten betreffen. Mehrere Detektionsalgorithmen werden auf Ihre Anwendbarkeit unter den gegebenen Messbedingungen untersucht. Dies umfasst Detektionsalgorithmen, die für die Detektion vollständig bedeckter Pixel optimiert sind wie die Normalized Cross Correlation (NCC), den Matched Filter Detektor (MF) und die Constrained Energy Minimization (CEM). Darüber hinaus werden die weit verbreiteten Subpixelfähigen Detektoren Adaptive Coherence / Cosine Estimator (ACE) und Adaptive Matched Subspace Detector (AMSD) betrachtet.
Der letztgenannte Detektionsalgorithmus basiert auf der strukturierten Variante des LMM, das mögliche Hintergrundspektren explizit anhand der enthaltenen Endmember modelliert. Diese müssen in einem generischen Messumfeld aus den erfassten Messdaten extrahiert werden. In dieser Arbeit wird mit dem Adaptive Background Generation Process (ABGP) ein zweistufiger Algorithmus zur Extraktion der Hintergrundspektren vorgeschlagen, der an die spezifischen Randbedingungen des Messsystems angepasst wurde.
Ein weiterer Parameter, der hierfür aus den gewonnen Messdaten ermittelt werden muss, ist die erwartete Modellordnung, die im Wesentlichen von der Anzahl der spektral verschiedenen Materialien in der Szene dominiert wird und die Anzahl der zu extrahierenden Endmember festlegt. In dieser Arbeit wird eine Methode zur Schätzung der Modellordnung vorgestellt, die auf dem Prinzip der Minimum Description Length (MDL) beruht. Dazu wurde die von Wax und Kailath vorgeschlagene MDL Methode zur Schätzung der Anzahl unabhängiger Signalquellen in komplexwertigen Radarsignalen auf die üblicherweise reellwertigen Signale in der Hyperspek-

tralbildverarbeitung adaptiert. Obgleich die Methode im Kontext der Hyperspektralbildanalyse entwickelt wurde, lässt sie sich direkt auf jede Form von Multi-Band Signale übertragen, die sich als lineare Superposition statistisch unabhängiger Signalquellen auffassen lassen.
Es wird gezeigt, dass die vorgeschlagene Methode, auch im Vergleich zu konkurrierenden Algorithmen zur Schätzung der Modellordnung wie die Noise Subspace Projection (NSP) oder Second Moment Linear Dimension (SML), in Simulationen sehr gute Ergebnisse auf Basis künstlich erzeugter Hyperspektralbilder liefert. Diese Ergebnisse lassen sich jedoch nicht auf Hyperspektralbilder übertragen, die mit der vorgeschlagenen Spektroskopie Methode gemessen wurden. Eine Analyse der Kovarianzmatrix-Eigenwerte legt den Schluss nahe, dass sich dies auf Speckle Rauschen zurückführen lässt. Speckle – ein multiplikatives, quasideterministisches Rauschen – wird durch die Kohärenz der Beleuchtung erzeugt und im Hyperspektralbildsensor zwar mittels eines Multi-Look Verfahrens reduziert, aber nicht vollständig eliminiert.
Die vorgeschlagene Methode zur Extraktion der Hintergrundspektren zeigt sich jedoch unempfindlich gegen eine überschätzte Modellordnung. Basierend auf diesem konnte daher der AMSD Algorithmus erfolgreich zur Detektion von Spuren aller erwähnten Explosivstoffe eingesetzt werden. Dies wird sowohl für die Detektion in Hyperspektralbildern gezeigt, die über kurze Distanzen von ca. 1.4 m gemessen wurden, als auch über lange Messdistanzen von bis zu 20 m.
Abschließend wird eine echtzeitfähige Variante der Laserrückstreuspektroskopie vorgestellt, die auf einem schnell abstimmenden EC-QCL beruht. Dabei wird die Emissionswellenlänge des Lasers mittels eines Mikro-Optoelektromechnischen Systems (MOEMS) innerhalb von 1 ms über den vollständigen Abstimmbereich von 7.5 µm bis 9 µm durchgestimmt. Dies ermöglicht eine Aufnahmegeschwindigkeit von bis zu 1 kHz von Rückstreuspektren mit einer Auflösung von 1 cm^{-1}. Dazu wird das diffus rückgestreute Licht von einem MCT Einzelelementdetektor eingesammelt um

– im Gegensatz zu Hyperspektralbildern – sequenziell Einzelspektren zu generieren. Es konnte experimentell demonstriert werden, dass sich die im Rahmen dieser Arbeit entwickelten Algorithmen zur Hyperspektraldatenanalyse für die Materialdetektion in Echtzeit eignen.

Abstract

Stand-off detection of hazardous substances has been subject to extensive research in the past years and – though several approaches have been proposed – remains to be an ambitious goal. Among the suggested methods, laser based measurement techniques have emerged as the most promising. In this work we present a stand-off long-wave infrared (LWIR) spectroscopy system for remote detection of hazardous substances. The principle is based upon wavelength selective illumination using External Cavity Quantum Cascade Lasers (EC-QCLs), that are tunable in the LWIR wavelength range from 7.5 µm to 10 µm. This spectral range is part of the molecular fingerprint region for many chemical compounds, including explosive substances like PETN, RDX, TNT and AN.

In the imaging variant of the hyperspectral data acquisition system, the diffusely backscattered light is collected by a high performance Mercury Cadmium Telluride (MCT) camera. Using synchronous tuning of the illumination wavelength, a hyperspectral image is created, where each pixel vector comprises the backscattering spectrum of a specific location in the scene. Every pixel vector is regarded as to be a linear superposition of few contributing spectra (endmembers) with unknown mixture weights. These endmembers are usually the spectra of the spectrally distinct materials in the scene and possibly one or several target substances of interest. This is mathematically described by the Linear Mixture Model (LMM), on which many existing hyperspectral target detection algorithms are based on.

In this work, we present details of the hyperspectral imaging sensor that touch both optimization of illumination as well as data acquisition and analysis. We analyze a variety of well known target detection algorithms for the task of substance detection in the acquired hyperspectral image data. The considered target detection algorithms include fast and intuitive full-pixel detection algorithms like the NCC, Matched Filter (MF) and CEM, as well as well known and powerful sub-pixel detection algorithms like the ACE and AMSD. The latter utilizes the structured LMM variant, that requires explicit knowledge of the contributing background materials in the scene. As these are in general unknown the corresponding endmembers have to be extracted from the available hyperspectral observation data using an endmember extraction algorithm. In this work, we present an endmember extraction algorithm, tailored to the boundary conditions, induced by the measurement technique.

An additional parameter that is required for data analysis is the number of spectrally distinct materials in the scene, that essentially comprises the required model order. We present a model order estimation method based on the Minimum Description Length (MDL) principle by adaption of a method proposed by Wax and Kailath [63] for estimation of the number of signal sources in complex radar observations. Whereas the method is developed in the context of hyperspectral image analysis, it can be readily applied to signal source estimation in any real-valued multi-band observations that comprise a linear superposition of independent sources.

Whereas the proposed MDL model order estimation method competes well or even outperforms comparable model order estimation algorithms like the NSP and SML on artificial data that was explicitly generated using the LMM, it significantly overestimates the required model order when applied to real-world measurement data. Based on observations of the corresponding covariance matrix eigenvalue-distributions, this behavior is attributed to correlated noise, most likely due to remaining speckle. The latter is caused by the coherent nature of the illumination source and significantly reduced – though not fully suppressed – using a multi-look approach.

The proposed background endmember extraction algorithm shows however, to be robust against model-overestimation. Based on the estimate of the background spectra, the AMSD algorithm is successfully applied for detection of all of the mentioned explosive substance residues on various substrates. This is demonstrated, both for hyperspectral image measurement results obtained by a short-range variant of the hyperspectral image sensor over 1.4 m, as well as with an extended range setup, operational up to distances of $\approx$ 20 m.

Finally, we present an extension of the backscattering spectroscopy method to a real-time measurement device, based on a rapid wavelength scanning EC-QCL. The latter swipes the full spectral emission range of the QCL chip within 1 ms which enables a spectral acquisition rate of 1 kHz. The diffusely backscattered light is in this measurement setup collected by a single-element MCT detector generating single spectra, rather than full hyperspectral images. We present an experimental setup capable of fast data acquisition and show, that the developed target detection algorithms are capable of real-time detection in the observed hyperspectral data.

Acknowledgements

I sincerely thank all of the many people who contributed to this work, in particular:

My supervisor Prof. Dr.-Ing. Jürgen Beyerer, for his expert opinions and challenging scientific questions and suggestions.

Prof. Dr. rer. nat. Oliver Ambacher, for his support and prudent advice.

Prof. Dr. Joachim Wagner, for his kind patience, his expert suggestions and continuous support.

Dr. Frank Fuchs, for his help, his trust, his support and for having made this work possible.

Dr. Stefan Hugger, who's help and contribution can not be overstated.

Dr. Marko Haertelt, for his support, suggestions and advice.

My colleagues of the workgroup *Laser Messtechnik*.

My co-workers in the department of Optoelectronics at IAF.

My mother, for her support and the judicious advice she offered while she was with us.

Elena, for her continuous indulgence and overwhelming support and my children Annabelle and Julian, for being an amazing source of joy and happiness.

Contents

1 Introduction

Within the past years, incidents involving bombs based on homemade explosives have emerged to be an increasing menace to the public. Instructions for producing highly effective explosive substances like Ammonium Nitrate / Fuel Oil (ANFO) or Tri-cyclic Acetone Peroxide (TATP) are easily accessible in the internet and can be prepared using common household appliances. In addition, many of the required ingredients are not regulated and thus can be purchased in public stores.
This has made e. g. ANFO and TATP frequent choices for explosive devices manufactured by terrorist organizations like Al-Qaida and ISIS. Furthermore, as no relevant acquisition network is required, extremists like Anders Behring Breivik (Oslo 2011) have been able to commit devastating strikes in public sites using home made explosive material. Politically or religiously motivated attacks might in some cases be preventable by appropriate surveillance of suspicious groups or organizations. However, in the past years a subculture of bomb builders has developed that are committed to manufacturing explosive devices as a hobby without planning to use them as weapons. Nevertheless, these illicit bomb-factories comprise a significant danger to public safety and – as there is no imminent motive – are extremely hard to be identified by security forces. This has raised public demand for systems that are capable of remotely detecting residues of dangerous substances.
Searching for and identifying an illicit bomb-factory is a difficult task for police and security forces. Intelligence data might often lead to surveillance of a specific object like a flat or van equipped with the required appliances.

It is then the next step to collect evidence that serves as reasonable ground for the suspicion. Nevertheless, it might be important to collect the evidence unobserved by the suspects. In this case a mobile, remote detection measurement device capable of trace detection of explosives across some ten meters would be of valuable assistance for security forces.
Airports and airplanes have repeatedly been subject to terrorist attacks within the last few years. While luggage and passengers are carefully scanned for metal objects, existing explosive substance detection techniques are applied only at random, as they require time-consuming swabbing. A measurement device that is capable of remotely detecting explosive residues and can be included into existing security routines with a small footprint could therefore strongly improve air travel safety.
Mobile security force units that secure an illicit bomb factory are often forced to operate under unclear and potentially life-threatening conditions. In general the hazardousness of chemical agents in the field can only be assessed using measurement techniques that require physical collection of the substance to be analyzed. Such units could highly benefit from a lightweight, mobile and real-time measurement device for non-contact chemical substance identification.
For successful operation under real-world conditions all applications mentioned require high true positive and low false alarm rates. In addition, the system should be able to operate unperceived and must not pose a threat in uncontrolled environments itself (e. g. by operating high power lasers). Boundary conditions that are likely to be relevant include measurement speed, mobility and system robustness.
In spite of extensive research effort within the last decade, remote trace detection of hazardous substances remains to comprise an ambitious goal, to which laser based methods proved to be among the most promising approaches. Van Neste et al. proposed a trace explosive detection system based on photoacoustic spectroscopy using quantum cascade lasers [61]. Kendziora et al. follow an approach called *photothermal infrared imaging*

spectroscopy that also employs QCLs for spectrally selective illumination [35]. Östmark et al. developed a stand-off imaging Raman spectroscopy device for detection of explosive residues [48], whereas Gottfried et al. employ laser induced breakdown spectroscopy for the same task [23]. To this day, none of these approaches were able to solve the task, matching all of the given conditions.

In this work we present a stand-off hyperspectral imaging technique for trace detection of explosives based on tunable infrared laser backscattering spectroscopy that showed promising results. Tunable long-wave infrared (LWIR) Quantum Cascade Lasers (QCLs) serve as wavelength selective illumination sources that cover the molecular fingerprint region between 7.5 µm and 10 µm. A high performance infrared camera is used to collect the diffusely backscattered light, giving rise to a hyperspectral image, where every pixel vector contains the backscattering spectrum of a specific location in the scene. Hyperspectral target detection algorithms are then applied to detect and identify known backscattering spectra of harmful substances..

We present two variants of the remote spectroscopy system: a short range hyperspectral image sensor for trace detection of substances at distances of up to 5 m and a long-range image sensor that proved successful for detection distances up to 20 m. Finally we present a mobile spectroscopy device based upon the same infrared laser backscattering spectroscopy system that enables non-contact, real-time measurements using rapidly tunable quantum cascade lasers as illumination source.

While much of the research effort was motivated by security related applications as mentioned above, the principle is by no means restricted to this field. The method can be readily applied to stand-off surface spectroscopy of any solid substance that shows spectral features in the considered wavelength range. This opens the door to a vast field of applications including food processing, industrial process control in the pharmaceutical industry or chemical surface analysis in materials processing.

This thesis covers the design of the hyperspectral image sensor as well as implementation of data acquisition software and algorithms for image processing, as combined research effort in these fields is required to achieve sufficient target detection performance. Particular focus is put on model driven hyperspectral target detection methods.
Two steps are common to many model driven target detection algorithms: selection of a model order and estimation of the probability distribution parameters for the test statistics. Whereas the latter is well covered by a variety of target detection algorithms, model order estimation has received less thorough attention by the hyperspectral image analysis community. Following the work of Wax and Kailath [63], we present a parameter-free algorithm towards model order selection, which is based on the well-known Minimum Description Length (MDL) principle. We evaluate the proposed method for application in hyperspectral images that were acquired using the proposed measurment method.
We present a detection performance comparison of several well known target detection methods on simulated data and analyze the impact of various noise sources on the expected performance and give several trace detection results of real-world measurements. Finally, an extension towards a real-time capable spectroscopy device is presented.

1.1 Contribution

The main objective of this dissertation is optimization of the proposed measurement method and development of hyperspectral image analysis algorithms for the task of remote trace substance detection. We aim for a hyperspectral image sensor that is capable of detection of a known set of hazardous substance residues on an a priori unknown background material with high probability of detection and low false-alarm rate. Thus, data acquisition and analysis have to be adapted to observed noise sources and characteristics that are specific to the proposed measurement technique.

The contribution of this dissertation may be summarized as follows:

Image acquisition and data pre-processing

The proposed method for hyperspectral image acquisition requires careful implementation of data-acquisition software and pre-processing steps. The former includes e.g. correct synchronization of image acquisition and laser emission wavelength tuning. During the pre-processing step, the observed raw image data requires correction of e.g. atmospheric transmission effects and noise induced by thin-film interferences, which is caused by the coherent nature of the illumination source.

Background subspace estimation

The target detection algorithm that is predominantly employed in this work is based on the structured variant of the Linear Mixture Model (LMM). I.e. the image background is explicitly described by the pure material spectra that contribute to the observed signal. This requires an endmember extraction algorithm that serves to generate the background subspace from the available image data. With the Adaptive Background Generation Process (ABGP) we propose a two-step approach to endmember extraction that is based upon the Adaptive Target Generation Process (ATGP) algorithm.

Evaluation of model order selection

As mentioned before, the data-analysis algorithms employed in this work are based upon the LMM, which explicitly describes the background in terms of the contributing material spectra that span the background subspace. We present and evaluate the performance of a method that aims to estimate the required number of dimensions of this subspace. The method is based on the MDL principle and is obtained by adaption of a method proposed for analysis of radar signals.

Evaluation of target detection framework

We present an analysis of measurement specific operation conditions (e.g. beam shape, spectral resolution, noise level, etc.) on the expected target detection performance, achieved by various data analysis algorithms. Finally, we present measurement results obtained on real-world samples containing various explosives and precursors (AN, PETN, RDX and TNT)

Extension to real-time sensing

We present an extension to real-time sensing based on the measurement principle. Real-time capability is achieved using a single element detector instead of a camera. Fast spectral scanning of the EC-QCL is achieved by replacing the mechanically tuned grating in the external cavity of the laser resonator by a Micro Opto Electromechanical System (MOEMS) element. In this setup a full spectrum is acquired within 1 ms, which renders the measurement principle applicable in dynamic scenes.

1.2 Structure of this Work

We introduce the principle of tunable laser backscattering spectroscopy together with the external cavity tuned quantum cascade laser in chapter 2. The mathematical methods and target detection algorithms are described in chapter 3. In chapter 4 we analyze the problem of model order selection in hyperspectral images and present an approach based on minimum description length. An overview over the existing approaches towards explosive substance detection is given in chapter 5. The short- and long-range hyperspectral imaging sensor setups are described in chapter 6. A detailed analysis of performance and real-world measurement results are presented in chapter 7. Results of real-time sensing with fast backscattering spectroscopy are given in chapter 8. We close with a summary and an outlook to further research in chapter 9.

2 Fundamentals of MIR Laser Spectroscopy

In this chapter, we follow Colthup et al. [13] and introduce the fundamentals of infrared laser spectroscopy as are relevant within the scope of this work. We give a brief introduction on External Cavity Quantum Cascade Lasers (EC-QCLs) that comprise the system's core laser source. Section 2.4 presents the concept of tunable laser backscattering spectroscopy for hyperspectral image acquisition. We close with a comment on speckle as noise source, caused by the coherent nature of illumination.

2.1 Infrared Fingerprint Spectroscopy

The mid-infrared (MIR) wavelength range from 2.5 µm to 25 µm is an especially interesting wavelength range for many spectroscopy applications, as most molecules show highly characteristic spectral absorption behavior in this range. It is therefore often referred to as *molecular fingerprint region* of the electromagnetic spectrum. Hence, several infrared spectroscopy methods have been developed that exploit the MIR range for various chemical analysis applications.

A radiation source is called *monochromatic*, if the frequency domain representation of the emitted light wave takes the form of a Dirac function $\delta(\lambda_p)$. In a particle view of light, the radiation source generates photons of wavelength λ_p that carries the energy $E_p = hc/\lambda_p$. If this photon energy is *absorbed* by a molecule, its energy – i. e. rotational, vibrational and electronic energy – is changed by the same amount $\Delta E_m = E_p$.

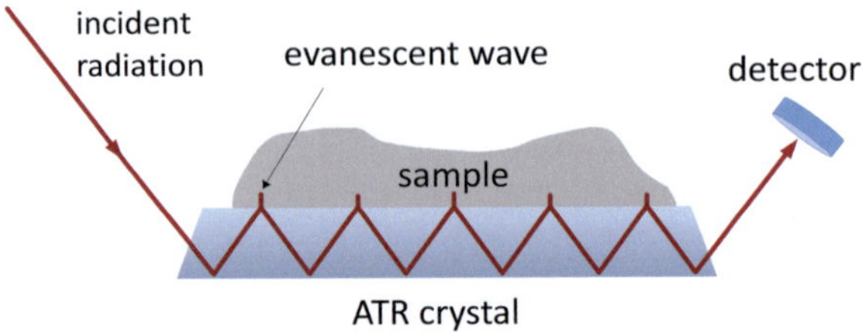

Figure 2.1: The principle of attenuated total reflection. The incident radiation travels through a crystal and is multiply reflected from the surface on which the sample is deposited. The evanescent wave at the points of reflection interacts with the sample, leading to a substance dependent attenuation of intensity.

A basic understanding of the absorption process can be supported by adopting a mechanical model of molecular structure, where the inter-nuclear forces are represented by mass-less springs and the nuclei are represented by point masses. In this model a molecule's internal motional degree of freedom is limited to specific modes of vibration and rotation. Incident monochromatic radiation translates into oscillatory mechanical force of defined frequency upon the nuclei. Depending on the number and masses of nuclei involved, as well as the spring-coefficients of their connecting springs, the resulting coupled oscillation is subject to resonances – i. e. certain excitation frequencies are more efficient in changing the vibrational mode level and hence the molecule's energy. Adopting a particle view of light again, photons of wavelengths that match the resonance frequencies are more likely to be absorbed by the molecule than photons of other wavelengths, where the interaction is weaker. Since the distribution of resonance frequencies is dependent on the molecular structure, analysis of the molecule's absorption wavelengths is highly substance specific and allows for chemical analysis.

In this work our main interest lies in spectroscopy and detection of explosive substances in the solid phase. Among the variety of measurement techniques available for solid phase substances, an important and well established method is internal reflection spectroscopy (IRS) or Attenuated Total Reflectance (ATR) [13]. The principle is illustrated in figure 2.1. Monochromatic infrared light travels through a crystal with refractive index n_1, such that the light wave is multiply reflected from the crystal surface on which the liquid or solid sample is deposited. The wave persists beyond the interface of the surface and thus reaches into the sample medium. The penetration depth into the sample medium d_p of this evanescent wave is determined by the wavelength λ, angle of incidence to the surface θ, and the refractive index of the surrounding medium n_2 (usually air, hence $n_2 = 1.0$):

$$d_p = \frac{\lambda}{2\pi n_1 \sqrt{\sin^2\theta - (n_2/n_1)^2}} \tag{2.1}$$

The evanescent wave interacts with the sample material, leading to attenuation of the total intensity. The attenuation of the incident power serves as measurement signal at the rear facet of the crystal.

Figure 2.2 shows infrared ATR spectra of solid explosive substances in the MIR wavelength range between 850 cm^{-1} and 4000 cm^{-1}. All substances show particularly pronounced absorption bands in the range between 850 cm^{-1} and 1600 cm^{-1}. Additional features can be observed in the spectral range around 3000 cm^{-1}, showing however considerably smaller cross sections. In addition, we show the characteristic atmospheric transmission at a range of 8 m in figure 2.2. It can be observed that in several wavelength ranges, severe absorption, predominantly caused by water in the atmosphere is expected. This is of major importance, considering the target application of stand-off spectroscopy, especially over long distances. Based on the atmospheric transmission spectrum in figure 2.2 in combination with the previous considerations on spectral feature distribution, we conclude that the most interesting measurement range in the MIR is the long-wave region between 850 cm^{-1} and 1330 cm^{-1}.

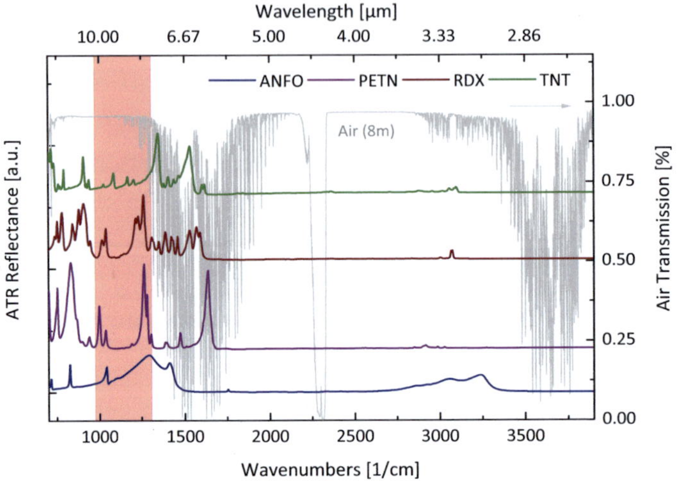

Figure 2.2: Typical reflectance spectra of explosive substances. The spectra were measured using ATR spectroscopy. All substances show most characteristic spectral features in the wavelength range between 800 cm^{-1} and 1500 cm^{-1}. In addition, the absorption spectrum of air shows a high transmittance within the measurement region.

We note that infrared spectra of smaller molecules that are of interest in many gas spectroscopy applications, typically show narrow-band absorption lines, whereas the resonance modes of more complex molecules give rise to the broader absorption features that are obtained in spectroscopy of solids. This effect can be observed in figure 2.2: the atmospheric absorptions caused by water are sharp in comparison to the spectral features of the explosives. Hence, the demands on spectroscopic laser sources for the two application are considerably different. While gas spectroscopy lasers are usually gas-specific and require narrow-band emission for the reasons stated, this demand is considerably relaxed for spectroscopy of solids. However, the latter application – especially if designed to be specific for a wide range of substances – requires far broader spectral tuning capability.

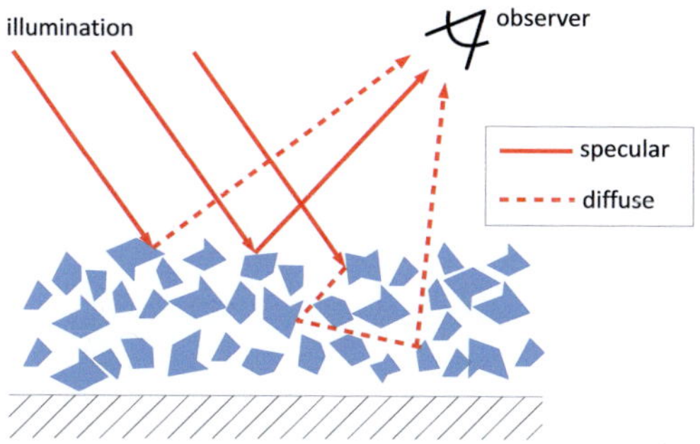

Figure 2.3: The measurement signal in diffuse reflectance spectroscopy of powders contains contributions of multiple internal absorption and scattering as well as specular reflections at the surface.

2.1.1 Comparison of Reflectance and ATR Spectra

Whereas ATR is a widespread and well established spectroscopy technique in many chemical analysis applications, it is for obvious reasons however not applicable to the stand-off geometry that is aimed for in this work. The laser backscattering technique employed in this work is based on diffuse reflectance spectroscopy using tunable infrared laser illumination. Among the established spectroscopy techniques, this is best comparable to Fourier Transform Infrared (FTIR) diffuse reflectance spectroscopy.

The measurement values obtained by ATR spectroscopy can be directly related to molecular absorption effects. This is no longer true for diffuse FTIR reflectance spectra of powders, as the diffuse reflectance of monochromatic incident radiation contains contributions of multiple internal absorption and scattering as well as specular reflections at the surface (figure 2.3), to which no general quantitative solutions exists [59].

Several phenomenological theories have been developed that express the reflectance in terms of absorption and scattering coefficients. The Kubelka-Munk theory [59] is frequently employed for diffuse FTIR reflectance spectroscopy. Assuming the sample is opaque – i. e. absorption of the sample is large enough to prevent a second scattering contribution at the substrate at all considered wavelengths – the Kubelka-Munk theory states that

$$\frac{K}{S} = \frac{(1 - R_\infty)^2}{R_\infty}, \tag{2.2}$$

where K denotes the absorption coefficient, S the scattering coefficient and R_∞ the (absolute) reflectance (the subscript ∞ indicates opaqueness of the sample). In general, both K and S can be wavelength dependent and hence, the reflectance spectrum matches the absorption spectrum, only if the scattering coefficient is constant.

Figure 2.4 illustrates the difference between the observed spectra in the wavelength range of interest. The ammonium nitrate ATR and diffuse reflectance spectra are well comparable, indicating that scattering is fairly constant in this case. In contrast, the ATR reflectance peaks are inverted in the diffuse reflectance spectrum for the case of TNT. The PETN and RDX diffuse reflectance spectra show mixtures of common and inverted peaks as well as derivation like features that can be attributed to predominantly specular surface reflections [65]. A thorough theoretical analysis of the physics causing such effects can be found in [65].

At this point, we adopt a phenomenological position and argue that the measurement values from diffuse reflectance spectroscopy might not be caused by molecular absorbance effects alone; however, the absorbance coefficient is involved causing the resulting spectra to be qualitatively at least as discriminative as ATR spectra. Other effects might add to the resulting signal but can be neglected, as long as they cause no unwanted dependencies of e. g. illumination or observation angle. We will employ empirical measurement results to support this assumption. We note however that this approach comes at the loss of quantitative analysis capability as well as a meaningful measurement unit and will use the term backscattering intensity rather than reflectance to support this notion.

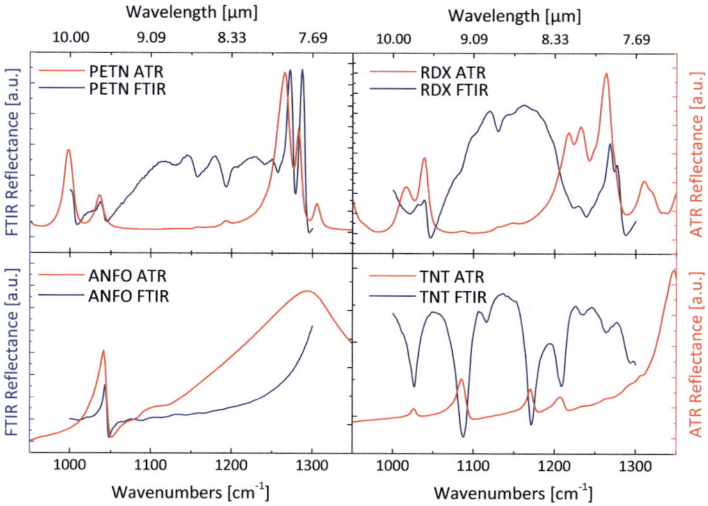

Figure 2.4: Comparison of spectra, obtained by ATR and FTIR diffuse reflectance spectroscopy. While some spectral features are observed by both measurement methods, others are inverted or show a derivation like appearance.

2.2 External Cavity Quantum Cascade Lasers for Infrared Spectroscopy Applications

The core element of the spectroscopy devices presented within this work is the Quantum Cascade Laser (QCL) [8]. The first QCL having been experimentally demonstrated in 1994 by Faist, Capasso, Sivco, Sirtori, Hutchinson and Cho, QCLs have since emerged as versatile tunable lasers for a wide range of mid-infrared spectroscopy applications. Whereas QCLs emitting in the THz regime [1, 62] as well as in the mid-infrared between 3.3 µm and 12 µm have been reported, main interest of this work lies in the spectral region between 7.5 µm and 11 µm. QCLs in this wavelength range are InGaAs/InAlAs semiconductor devices grown on InP substrates using molecular beam epitaxy. The emission wavelength range can be tailored to the application by band-structure and quantum mechanical wavefunction engineering.

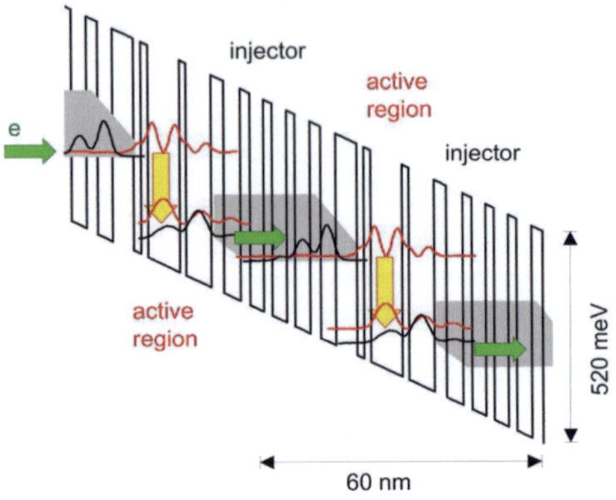

Figure 2.5: Energy diagram that illustrates the principle of a quantum cascade laser (from [8])

2.2.1 The Basic QCL Principle

In conventional bipolar diode lasers, photons are generated by radiative recombination of electrons and holes. Hence, the emission wavelength is directly dependent upon the energy gap between valence band and conduction band of the material system.

In contrast, QCLs generate photons by electrons that undergo transitions between quantized energy levels within the conduction band. Quantum wells are artificially created in the material system by band structure engineering and define the electrons' allowed quantized energy levels – and consequently the emission wavelength – independent of the material system's intrinsic energy gap.

Electrons enter the active region composed of several quantum wells via an electron doped injection region. Assuming the electron's initial energy level is E_n, relaxation into the next lower energy level will be E_{n-1} and a photon of wavelength $h/(E_n - E_{n-1})$ will be emitted, with h denoting Planck's constant. As the new energy state is still within the conduction band, the electron can be employed for a second radiative recombination, by feeding it into a subsequent active region. Typical QCLs contain 20 to 35 such stages.

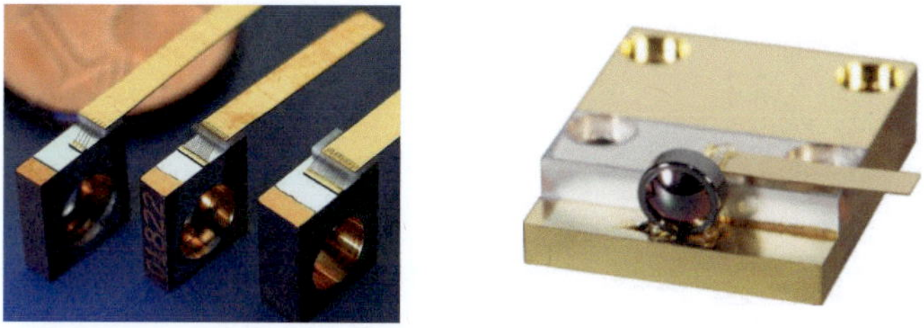

Figure 2.6: Typical QCL chips that were designed, grown and processed at Fraunhofer IAF. The chips are soldered on gold plated heat sinks using Indium-solder for optimized heat dissipation.

Figure 2.5 (left) shows the conduction band profile and quantized energy states of a QCL designed to emit at $\lambda = 7.5\,\mu\text{m}$, illustrating the principle. The radiative transitions takes place between energy levels three and two. To enable lasing the corresponding population inversion time has to be substantially longer than the lifetime of level two. To achieve this, the lowest energy level is placed just below the lower laser level with a longitudinal optical phonon energy spacing, causing electrons in the latter state to be rapidly scattered in a non-radiative transition.

Using band-structure and wave-function engineering QCLs can be designed to offer broad spectral tuning capability. This can be achieved, by splitting the lower laser energy level into several closely spaced sub-levels, hence spreading possible energy transitions in the quantum well. This approach is called *bound-to-continuum design* [14]. In contrast, *heterocascading* QCLs achieve broadband tuning capability by combination of different wavelength centered active regions within a single chip [44].

Figure 2.6 shows typical QCL chips that were designed, epitaxially grown and processed at Fraunhofer IAF. The laser chips are Indium-soldered on gold coated heat sinks.

2.2.2 External Cavity QCL Tuning

QCL chips can be operated as Fabry-Pérot chips by coating the rear chip facet with a high reflectivity coating and the front facet with a partial anti-reflective coating. In this configuration the chip facets define a laser resonator that allows lasing at several longitudinal modes. Hence, Fabry-Pérot QCL chips in general show a broad emission spectrum. However, for QCL operation in spectroscopy applications it is eminent to gain control over the emission wavelength.

A common way of achieving narrow-band lasing at a specific wavelength is the concept of distributed feedback (DFB) lasers, where a periodic grating structure is processed in an additional material layer on the active region of the laser chip. The periodic structure acts as a diffraction grating and the grating period defines the resulting emission wavelength. Wavelength tuning of DFB QCLs can be achieved to some extent by manipulating the chip temperature, which alters the refractive index and consequently the effective grating period. Typical tuning ranges from $2\,cm^{-1}$ to $3\,cm^{-1}$ are achievable, if heating by injection current is applied. This can be extended to as much as $10\,cm^{-1}$ to $20\,cm^{-1}$, if slower external temperature control methods are employed [64]. Thermal tuning approaches are frequently employed in gas spectroscopy applications (e. g. [6]). However, due to the spectral width of the characteristic absorption features, the tuning capability achievable using such thermal effects – while acceptable in many gas sensing applications – is insufficient for spectroscopy of solids.

Lee et al. developed a DFB QCL array that covers the wavelength range between 8.0 µm and 9.8 µm [39]. The array consists of 25 bound-to-continuum DFB-QCL chips with varying grating periods, such that the chips' emission wavelengths are $9.5\,cm^{-1}$ apart. This gap could be potentially filled using the aforementioned thermal tuning approach. A comprehensive overview over the impressive advances achieved in the field of DFB-QCL arrays is given by Rauter and Capasso in [50].

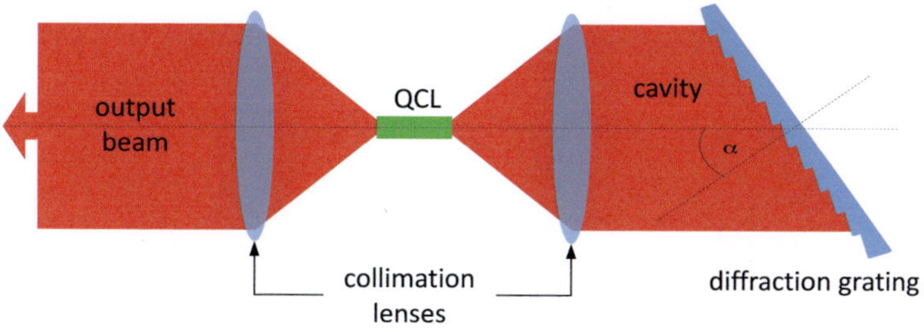

Figure 2.7: The principle of an external cavity tuned QCL using a blazed diffraction grating in Littrow configuration. The emission wavelength is selected by choice of the grating angle α that determines the wavelength which is fed back into the laser chip.

However, whereas the concept of thermal tuned DFB-QCL arrays potentially allows narrow-band tuning over a wide spectral range, it comes at the cost of extensive technological expense.

As an alternative, an external cavity resonator can be used to achieve broad-band wavelength tuning of QCLs that leads to the EC-QCL. Figure 2.7 illustrates the principle of an EC-QCL in a Littrow configuration. The rear laser facet is usually coated with an anti-reflection coating to make it transparent for incident light at the resonator side of the chip and a partially reflecting high reflectivity coating is applied to the chip's front facet. The light emitted from the rear laser facet is collimated and a blazed reflective diffraction grating serves as spectrally selective feed back element. In Littrow configuration the first diffraction order wavelength is reflected back into the laser chip, amplifying the corresponding lasing mode and giving rise to a standing wave between grating and front laser facet. The output wavelength is determined predominantly by the angle of the diffraction grating to the laser chip axes, and thus follows the grating equation:

$$\lambda_{out} = 2g\sin(\alpha) \tag{2.3}$$

where α denotes the grating angle and g denotes the grating constant.

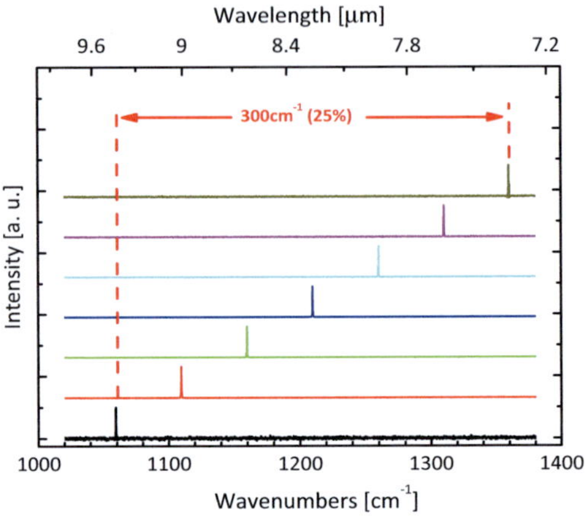

Figure 2.8: Typical tuning behavior of an external cavity tuned QCL. The emission wavelength of the room-temperature operated QCL chip is tunable from 1050 cm^{-1} to 1400 cm^{-1} using 100 ns pulses and a repetition rate of 10 kHz.

We note that equation 2.3 comprises an approximation that suffices for the target application, but ignores several physical effects that affect the emission wavelength and bandwidth [64]. The illuminated number of grating lines is finite, causing a blurring of the wavelength selection function. I. e. the wavelength selection function is not infinitely sharp but Gaussian shaped with a typical 2 σ width in the range of 2 cm^{-1} depending on the number of illuminated grating periods. This potentially causes several adjacent wavelengths to be amplified for a specific grating angle. Additionally, in the fixed length external cavity setup employed in this work, the standing laser wave is subject to the resonator's geometrical confinement that allows only specific modes to run – the *external modes*. Finally, the resonator facet's anti-reflection coating is usually not perfect, and thus gives rise to *internal* laser modes that are subject to the confinement defined by the geometry of the laser chip itself. If – as is the case in this work – the QCL is operated in pulsed mode, the chip temperature and hence the material's

refractive index shifts during the pulse. Depending on the dimension of the laser chip and the emission wavelength, several adjacent wavelengths may be excited consecutively. Typical QCL chips with an optical length of 1 cm show an internal mode spacing of $\approx 0.5\,cm^{-1}$ and two to three internal modes are excited within a single pulse. These operation conditions limit the effective average emission bandwidth to approximately $1\,cm^{-1}$.
Figure 2.8 shows lasing spectra of a typical external cavity QCL operated in Littrow configuration measured by FTIR spectroscopy [17]. By rotation of the grating, the emission wavelength of the considered QCL is tunable from $1050\,cm^{-1}$ to $1400\,cm^{-1}$ at room-temperature using 100 ns pulses and a repetition rate of 10 kHz.

2.3 Infrared Laser Backscattering Spectroscopy

The basic measurement principle of the remote sensing device presented within this work is tunable infrared laser based backscattering spectroscopy: the sample is illuminated by a wavelength tunable EC-QCL and the diffusely backscattered light is observed by a detector. The backscattering spectrum is obtained by measuring the backscattering intensity for several illumination wavelengths.
As noted in section 2.1.1, the diffuse reflectance of a non-isotropic, opaque sample, can be phenomenologically described by the Kubelka-Munk theory; i. e. the observed backscattering signal is considered a mixture of specular and multiple internal scattering, absorption and specular surface reflectance [59]. However, as we do not require the absolute absorption coefficients as obtained by ATR spectroscopy, we dismiss the associated additional complexity induced upon measurement conditions and data analysis. For the task of target detection it suffices to adopt the simpler model of Lambertian scattering [32]. This amounts to collecting the various signal contributions in the Kubelka-Munk theory (multiple internal scattering, absorption and specular surface reflection) into a single wavelength-

dependent coefficient that we showed to be substance-specific. Following Lambert's cosine law, the reflectance of an ideal Lambertian surface is independent of the observation angle. The measured backscattering intensity is predominantly a function of angle of incident illumination ϕ and the illumination wavelength λ. Let us assume for now, the illumination source is capable of infinitely narrow emission bandwidth and it's emission intensity at λ be $I_0[\lambda]$. Ignoring losses due to atmospheric absorption the observed intensity is then expected to be proportional to

$$I[\lambda] \sim I_0[\lambda] \cdot a_M[\lambda] \cdot \cos(\phi). \tag{2.4}$$

The coefficients $a_M[\lambda]$ denote wavelength dependent absorption effects related to the structure of molecules that comprise the illuminated material and are thus the magnitude of interest. The square brackets in equation 2.4 indicate that the wavelength λ is considered a discrete quantity. During a measurement – i. e. variation of λ – the parameter ϕ is considered constant. The illumination intensity $I_0(\lambda)$ is laser dependent and its impact on $I[\lambda]$ can be eliminated by a reference measurement of $I_0[\lambda]$. Similarly we account for atmospheric transmission variations $\tau_{Atmosphere}[\lambda]$ and obtain the measurement signal

$$a_M[\lambda] \sim \frac{I[\lambda]}{I_0[\lambda] \cdot \tau_{Atmosphere}[\lambda]} \tag{2.5}$$

that yields the backscattering spectrum

$$\vec{a} = [a_M[\lambda_1], \ldots, a_M[\lambda_p]]^\top. \tag{2.6}$$

Note that the measured backscattering intensities $a_M[\lambda]$ are in general subject to an unknown (wavelength independent) scaling factor.
This is for the case of an illumination source with infinitely narrow emission bandwidth. However, as stated before, pulsed EC-QCLs in Littrow configura-

tion are subject to various physical effects that cause emission bandwidth blurring as well as spectral mode-hopping. The expected spectral resolution of typical EC-QCLs in pulsed mode is thus expected to be $\approx 1\,\text{cm}^{-1}$. This determines $1\,\text{cm}^{-1}$ as the lower limit of achievable spectral resolution of a spectrometer based on such a laser. In addition, as the measured spectra can be considered to comprise convolutions of the true backscattering spectra with the impulse response of the laser emission spectrum, the multi-mode emission of the EC-QCL may be expected to cause difficulties, if the bandwidth of the spectral features is in the order of or even smaller than the laser emission bandwidth. Experimental data suggests however that due to the broad spectral features this effect can be neglected in spectroscopy of solids. This is supported by comparison of spectra obtained by laser backscattering spectroscopy and FTIR reflectance spectroscopy given in section 6.6. We hence consider the measurement vector $\vec{\mathrm{a}}$ in equation (2.6) an adequate approximation to the backscattering spectrum for the target application.

2.4 Active Imaging Spectroscopy

The hyperspectral image sensor presented within this work is based upon the principle of backscattering spectroscopy, introduced in the previous section 2.3. A tunable long-wave infrared (LWIR) EC-QCL laser illuminates the surface to be analyzed and – instead of a single element detector – an imaging sensor is used to capture the diffusely backscattered light (figure 2.9). Let the laser be tunable to the target wavelengths of interest denoted $\{\lambda_1, \ldots, \lambda_p\}$. To generate a hyperspectral image, the illumination wavelength is set to λ_1 and a spectrally broad image sensor collects the diffusely reflected light. We call this image $\mathrm{I}_{\text{active}}[\lambda_1]$.

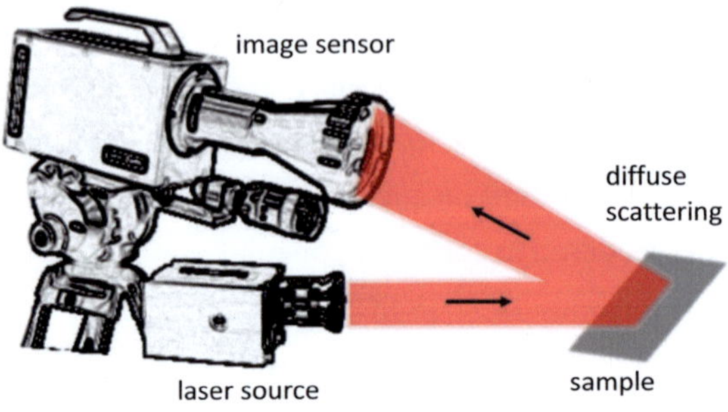

Figure 2.9: The general principle of backscattering spectroscopy using active, wavelength-selective illumination. A tunable laser is used to illuminate the scene, while an image sensor collects the diffusely backscattered light.

In a next step, the laser is switched off, and the thermal background image $\mathrm{I}_{\text{thermal}}[\lambda_1]$ is recorded. The difference image

$$\Delta\mathrm{I}[\lambda_1] = \mathrm{I}_{\text{active}}[\lambda_1] - \mathrm{I}_{\text{thermal}}[\lambda_1]$$

is calculated and forms the image layer corresponding to illumination band λ_1 of the resulting hyperspectral image. Repeating this process for all remaining target wavelengths λ_2 to λ_p generates the difference image stack

$$\mathrm{X} = \{\Delta\mathrm{I}[\lambda_1], \ldots, \Delta\mathrm{I}[\lambda_p]\}. \tag{2.7}$$

If the image sensor has M pixel rows and N pixel columns the matrix X has the dimension $\mathbb{R}^{M \times N \times p}$. We call X a hyperspectral image. The $M \cdot N$ vectors $\vec{\mathrm{x}}_i$ comprising the matrix X are the backscattering spectra of the corresponding locations of the illuminated scene.

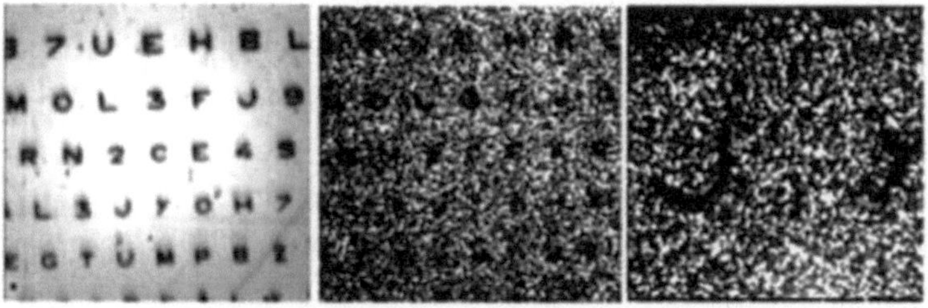

Figure 2.10: Example of speckle effect caused by coherent illumination source (from [22]). Left: Speckle-free image taken under incoherent light conditions. Center: the identical scene shows strong speckle under coherent illumination. Right: Magnification of a portion of the center image.

2.5 Laser Speckle and Beam Shaping

The coherent nature of the illumination source in the measurement geometry gives rise to a semi-statistical effect called *speckle* that is caused by the roughness of the illuminated surface. The statistical surface roughness induces a phase modulation on the wave-front of the laser wave, causing deterministic (i. e. reproducible) interference patterns at the detector – hence the term semi-statistical. Figure 2.10 shows the difference between imaging under incoherent (i. e. speckle-free) and coherent illumination conditions. As speckle is predominantly an effect of coherence, rather than wavelength, it has been subject to extensive research in various fields throughout the electromagnetic spectrum and several solutions towards speckle suppression have been proposed. The majority of such are based on the principle of averaging several different speckle patterns. Figure 2.11 illustrates an example of this approach using multiple illumination sources. Illumination with a single coherent illumination source generates a specific speckle pattern on the sensor. Adding additional spatially displaced coherent illumination sources will add corresponding speckle patterns and hence reduce the speckle noise effect, if the illumination sources are mutually incoherent.

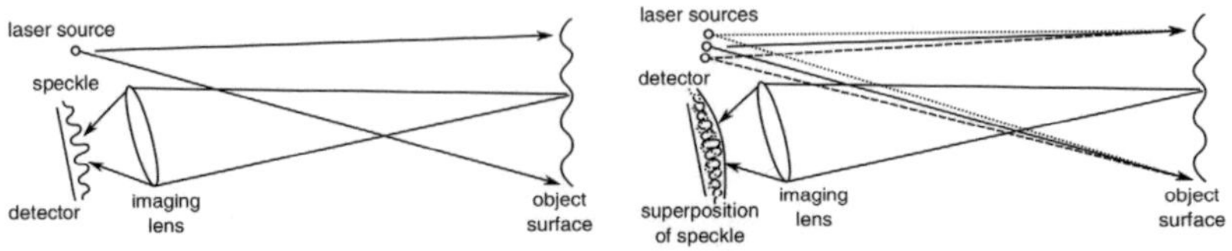

Figure 2.11: Left: illumination with a single coherent illumination source causes a specific speckle pattern on the detector. Right: speckle is reduced by a multi-source solution, if the coherent illumination sources are independent (source [38]).

We note that the latter condition is crucial for speckle reduction. If the sources are dependent, i. e. the incident laser waves have fixed phase relations, the additional illumination sources will alter the resulting speckle pattern while the speckle contrast is left unchanged.

The speckle noise reduction of a multi-pattern approach can be quantified using the statistics of speckle that were thoroughly investigated by Goodman. Following [22] speckle noise follows an exponential distribution and its effect can be measured in terms of the speckle contrast. Let I denote the irradiance caused by speckle. The probability distribution $p(I)$ is then given by

$$p(I) = \frac{1}{\mu_I} \cdot e^{-\frac{I}{\mu_I}} \tag{2.8}$$

where μ_I denotes the mean irradiance. Letting σ_I denote the variance, the speckle contrast is given by

$$C = \frac{\sigma_I}{\mu_I}. \tag{2.9}$$

Goodman concluded that averaging over M uncorrelated speckle patterns reduces the speckle contrast by a factor of $\sqrt{M}$

$$C_M = \frac{C}{\sqrt{M}}. \tag{2.10}$$

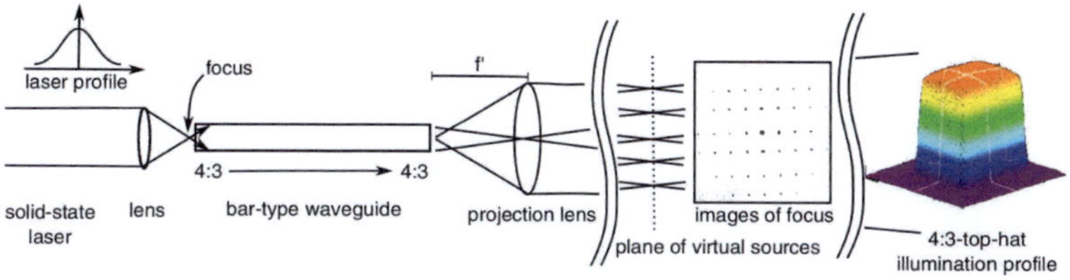

Figure 2.12: Laurenzis et al. suggested an approach to speckle reduction in active range-gated images based on a waveguide (from [38]).

Laurenzis et al. suggested an approach to speckle reduction in the short-wave infrared spectral range, based upon the aforementioned multiple source solution in [38]. A pulsed Nd:YAG pumped KTP Optical Parametric Oscillator (pulse length 9 ns) emitting at $\lambda = 1.57\,\mu\text{m}$ serves as radiation source and its output beam is focused on the facet of a rectangular waveguide. Traveling through the waveguide the light wave is subject to multiple internal reflections, each of which giving rise to a virtual illumination source, spatially displaced from the location of the true illumination source. A projection lens is used to project the intensity distribution at the exit facet onto the scene (figure 2.12). Given the specific waveguide and geometry, a speckle contrast reduction of $\sqrt{M} > 4.5$ is reported [38]. The significant reduction of laser speckle indicates that the virtual sources are incoherent among each other, which in consequence implies that the coherence length of the illumination source is shorter than the distance of the virtual sources. This notion is supported by the short pulse length.

Experiments showed that this approach is not directly transferable to speckle reduction in the mid-infrared range using QCLs as illumination source. This is attributed to the fact that the coherence length is significantly longer than the distance between the virtual sources. However, a modification of the approach of Laurenzis et al. has been employed, that enables speckle reduction to a level allowing application of most of the data analysis algorithms considered in this work.

3 Hyperspectral Image Processing

In this section we give a definition of a hyperspectral image and introduce the fundamental model of the hyperspectral image analysis algorithms that are presented in the following. Though there are other approaches for hyperspectral target detection based on e. g. machine learning, in this work the focus will be upon model based approaches.
Throughout this work scalars will be denoted by italic letters, vectors will be indicated by an arrow, and matrices by upper-case letters. Exceptions to this general rule will be indicated.

3.1 Definitions

Within the scope of this work a hyperspectral image is defined as a multi-channel image comprising p-dimensional real (or in rare cases complex) valued vectors $\vec{\mathrm{x}} \in \mathbb{R}^p$ that are arranged into a three-dimensional image cuboid X having M columns and L rows:

$$\mathrm{X} = \{\vec{\mathrm{x}}_{i,j} \,|\, \vec{\mathrm{x}} \in \mathbb{R}^p, i = 1 \dots L, j = 1 \dots M\} \tag{3.1}$$

In many cases, we will drop the notion of a three dimensional *image stack* in favor of a two dimensional *hyperspectral observation* denoted as

$$\mathrm{X} = \{\vec{\mathrm{x}}_i \,|\, \vec{\mathrm{x}} \in \mathbb{R}^p, i = 1 \dots N\}, \text{with} N = M \cdot L. \tag{3.2}$$

Any given hyperspectral image can be converted to match either definition by a simple rearrangement of the observation vectors $\vec{\mathrm{x}}_i$.

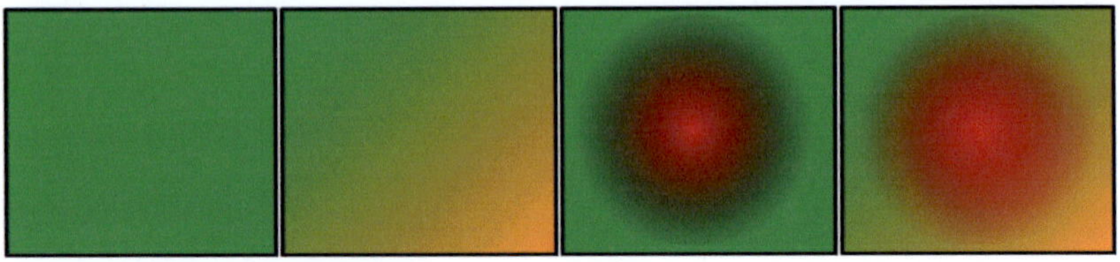

Figure 3.1: The linear mixture model assumes that a pixel contains either of (from left to right): homogeneous or heterogeneous background, target material in homogeneous or heterogeneous background.

The term *hyperspectral image* can thus be used interchangeably referring to either one of the definitions that is more suitable within the given context. Physically meaningful spectra are positive valued and hence it seems a natural choice to stipulate $\vec{x} \in \mathbb{R}^p_{\geq 0}$. Due to the measurement principle employed in this work that is based on difference images (section 2.4), negative values frequently occur in low signal regions of the hyperspectral image. It is thus in this case reasonable to adopt the more general model formulation given in equation 3.2.

3.2 The Linear Mixture Model

The fundamental model on which most model based approaches to hyperspectral image analysis are based upon, is the Linear Mixture Model (LMM). The principle is outlined in figure 3.1: the signal measured by a single camera pixel can either be homogeneous or heterogeneous background material, or target material embedded into a homogeneous or heterogeneous background. For an arbitrary pixel vector $\vec{x}_i$ we thus obtain the mathematical definition:

$$\vec{x}_i = S \cdot \vec{\alpha}_i + \vec{n}_i,\ S \in \mathbb{R}^{p \times q}, \vec{\alpha}_i \in \mathbb{R}^q, \vec{n}_i \in \mathbb{R}^p. \tag{3.3}$$

The matrix $S = [\vec{s}_1 \cdots \vec{s}_q]$ is often referred to as material or endmember matrix as it contains the set of spectra of the pure materials in the scene from which any other spectrum in the image can be generated by linear mixing. The elements of the weighting vector $\vec{\alpha}_i = [(\vec{\alpha}_i)_1 \cdots (\vec{\alpha}_i)_q]^\top$ are commonly known as abundances that comprise the mixture weights for the observation $\vec{x}_i$ and several target detection algorithms aim to exploit the physically motivated boundary condition

$$(\vec{\alpha}_i)_j \geq 0, \forall j = 1 \ldots q. \tag{3.4}$$

This condition is referred to as non-negativity constraint on the model abundances. A second common boundary condition often applicable to the weighting elements is the sum-to-one constraint, asserting that

$$\sum_{j=1}^{q} (\vec{\alpha}_i)_j = 1. \tag{3.5}$$

Whereas equation 3.5 implies the assumption that the spectra forming the spectral matrix are identically scaled to the data in the observation matrix X several target detection algorithms like the Fully Constrained Least Squares Detector (FLCS) and the Non-negativity Constrained Least Squares Detector (NCLS) have been proposed on the basis of such geometrical considerations [29]. The LMM is completed by the additive noise term $\vec{n}_i$ that accounts both for possible model errors, as well as for any noise sources induced by the measurement device, used to acquire the observation data.

3.3 Target Detection Algorithms

Target detection algorithms aim to detect the pixel vectors in a hyperspectral image that contain one or many target spectra. We call the set of target spectra *target library* and denote it as matrix $\mathrm{T} \in \mathbb{R}^{p \times t}$ containing the target spectra as columns, with t denoting the number of targets in the library. Three broader families of approaches can be identified within target detection algorithms.

Full pixel target detection algorithms

The family of *full pixel target detectors* can be distinguished by the property of sub-pixel target detection capability. Sub-pixel detection capability is given, if a target detection algorithm is able to detect target spectra that are mixed into the pixel vector under test given a specific mixture model. Full-pixel target detectors do not require specification of an underlying data model and are thus expected to yield best detection performance for detection of pure target material. A well known member of this family is the Normalized Cross Correlation (NCC) algorithm (section 3.3.2).
Sub-pixel target detectors are a refinement upon this approach as they aim to detect the target material in pixel vectors that are only partially covered by the target spectrum and contain spectra of the background material otherwise, thus requiring a mixing model. Most approaches choose the LMM given in section 3.2. Among the sub-pixel target detection algorithms two fundamentally different model based approaches can be identified.

Structured target detection algorithms

Structured target detection algorithms require knowledge of the background spectra in the hyperspectral image to be analyzed. These background spectra are typically collected as columns forming the background matrix $\mathrm{B} \in \mathbb{R}^{p \times b}$. The substance matrix $\mathrm{S} \in \mathbb{R}^{p \times q}$ in equation 3.3 is thus separated into the target matrix T and the background matrix B, which yields

$$\vec{\mathrm{x}} = \mathrm{T} \cdot \vec{\alpha}_{\mathrm{T}} + \mathrm{B} \cdot \vec{\alpha}_{\mathrm{B}} + \vec{\mathrm{n}}. \tag{3.6}$$

Equation 3.6 will often be referred to as *full model*. The estimated elements of the target abundance vector $\vec{\alpha}_{\mathrm{T}}$ and the background abundance vector $\vec{\alpha}_{\mathrm{B}}$ are then used to derive a binary classification result for each pixel vector and target spectrum. A famous member of this algorithm family is the Adaptive Matched Subspace Detector (AMSD) described in section 3.3.6.

Unstructured target detection algorithms

In contrast to structured target detectors, *unstructured* target detection algorithms refrain from explicitly characterizing the background subspace. Instead, the background is modeled as random noise clutter in which the target spectra are embedded. In mathematical terms, the linear mixture model reads:

$$\vec{\mathrm{x}} = \mathrm{T} \cdot \vec{\alpha}_{\mathrm{T}} + \vec{\mathrm{n}}. \tag{3.7}$$

The Adaptive Coherence / Cosine Estimator (ACE) described in section 3.3.7 is a well known target detection algorithm from this family.

3.3.1 Formalization

The considered target detection algorithms can be denoted and thought of as functions T_D that – given a target spectrum[1] $\vec{\mathrm{t}} \in \mathbb{R}^p$ – map the pixel vectors $\vec{\mathrm{x}}_i$ onto a scalar:

$$T_D : (\mathbb{R}^p; \vec{\mathrm{t}}) \rightarrow \mathbb{R} \tag{3.8}$$

Hence, applying a specific target detection algorithm upon the elements of a hyperspectral observation $\{\vec{\mathrm{x}}_i\}_{i=1}^N$ will produce a set of scalars $\{T_D(\vec{\mathrm{x}}_i; \vec{\mathrm{t}})\}_{i=1}^N$. The values of $T_D(\vec{\mathrm{x}}_i; \vec{\mathrm{t}})$ usually relate to the probability of the pixel vector $\vec{\mathrm{x}}_i$ containing the target spectrum $\vec{\mathrm{t}}$, with large values indicating a high contamination probability. However, for the task of target detection, it is desirable to establish a binary decision for each pixel vector, whether it contains a specific target or not. This can be achieved by choice of a threshold value $\tau \in \mathbb{R}$ that is applied to the detection function output.

[1] Some of the algorithms are defined to operate on target *matrices* T rather than target vectors, containing e. g. known spectral variants of the same target material as columns. In this case the detection functions take the form $T_D : (\mathbb{R}^p; \mathrm{T} \in \mathbb{R}^{p \times t}) \rightarrow \mathbb{R}$.

This leads to the target *classification function* $\Delta_D(\vec{x};\vec{t},\tau)$:

$$\Delta_D(\vec{x};\vec{t},\tau) = \begin{cases} 0, T_D(\vec{x};\vec{t}) <= \tau & \text{(background)} \\ 1, T_D(\vec{x};\vec{t}) > \tau & \text{(target present).} \end{cases} \tag{3.9}$$

The choice of τ determines a detection algorithm's True Positive Rate (TPR) and False Alarm Rate (FAR). Moreover, TPR and FAR will in general depend upon the noise variance and the background interference that might not be constant at different locations in the image – a highly undesirable effect in most practical applications. However, there are some algorithms for which this is not the case. These algorithms are said to have Constant False Alarm Rate (CFAR) property, as the FAR is independent of noise variance and background interference.

3.3.2 The Normalized Cross Correlation Detector

The Normalized Cross Correlation (NCC) target detector can be considered the most intuitive and straight-forward approach towards target detection. As it does not require a specific data model it belongs to the family of full-pixel target detection algorithms.
The detection function is given by the definition of normalized cross-correlation and reads:

$$T_{\mathrm{NCC}}(\vec{x};\vec{t}) = \frac{1}{p-1}\frac{(\vec{x}-\mu_{\vec{x}})^\top \cdot (\vec{t}-\mu_{\vec{t}})}{\gamma_{\vec{x}} \cdot \gamma_{\vec{t}}}, \tag{3.10}$$

where p denotes the number of bands, $\gamma_{\vec{x}}$ and $\gamma_{\vec{t}}$ denote the standard deviations of the pixel vectors under test and the target spectrum and their means are denoted $\mu_{\vec{x}}$ and $\mu_{\vec{t}}$. The normalization ratio $N-1$ refers to the sample size of the hyperspectral observation $\{\vec{x}_i\}_{i=1}^N$.

3.3.3 The Matched Filter Detector

The Matched Filter (MF) for hyperspectral target detection given e.g. in [41] is a widely used algorithm from the family of full-pixel detectors. Assuming multivariate normal distributions for background clutter and target spectrum it takes a statistical hypothesis testing approach to detection of a target $\vec{t}$ in the observation $\{\vec{x}_i\}_{i=1}^{N}$ to discriminate the two competing hypotheses:

$$\begin{aligned} H_0 &: \vec{x}_i = \vec{n}, \quad \vec{n} \sim \mathcal{N}(\vec{\mu}_B, \Gamma_B) \\ H_1 &: \vec{x}_i = \vec{t}, \quad \vec{t} \sim \mathcal{N}(\vec{\mu}_t, \Gamma_t). \end{aligned} \tag{3.11}$$

With $f(\cdot|\cdot)$ denoting the respective probability density functions, the Neyman-Pearson (NP) optimum likelihood-ratio test reads:

$$\mathrm{LRT}(\vec{x}_i) = \frac{f(\vec{x}_i|H_1)}{f(\vec{x}_i|H_0)}. \tag{3.12}$$

Adopting multivariate normal probability density functions in equation 3.12 and applying a logarithm this leads to the optimum-NP detector [33]:

$$T(\vec{x};\vec{t}) = (\vec{x} - \vec{\mu}_B)^\top {\Gamma_B}^{-1} (\vec{x} - \vec{\mu}_B) - (\vec{x} - \vec{\mu}_t)^\top {\Gamma_t}^{-1} (\vec{x} - \vec{\mu}_t) \tag{3.13}$$

In general the distribution parameters under the target hypothesis are not known a priori. Hence, it is assumed, that $\vec{\mu}_t = \vec{t}$ and that background and target covariance matrices are equal $\Gamma_B = \Gamma_t = \Gamma$. This allows equation 3.13 to be re-written to

$$T(\vec{x};\vec{\mu}_t) = \kappa^{-1} \cdot (\vec{t} - \vec{\mu}_B)^\top \Gamma^{-1} (\vec{x} - \vec{\mu}_B), \tag{3.14}$$

which is a scaled variant of the squared Mahalanobis distance from the pixel under test to the background distribution. In practical applications, the background distribution parameters mean vector $\vec{\mu}_B$ and covariance matrix

Γ_{B} are replaced by their Maximum Likelihood Estimates (MLEs) obtained from background training data and κ is set to

$$\kappa = \left(\vec{t} - \vec{\mu}_{\mathrm{B}}\right)^{\top} \Gamma^{-1} \left(\vec{t} - \vec{\mu}_{\mathrm{B}}\right), \tag{3.15}$$

which serves to normalize the detection output to one if $\vec{x} = \vec{t}$. This yields

$$T_{\mathrm{MF}}(\vec{x};\vec{t}) = \frac{(\vec{t} - \vec{\mu}_{\mathrm{B}})^{\top}\Gamma^{-1}(\vec{x} - \vec{\mu}_{\mathrm{B}})}{(\vec{t} - \vec{\mu}_{\mathrm{B}})^{\top}\Gamma^{-1}(\vec{t} - \vec{\mu}_{\mathrm{B}})} \tag{3.16}$$

as closed form MF target detection function.

3.3.4 The Constrained Energy Minimization

The Constrained Energy Minimization (CEM) originally proposed by Farrand and Harsanyi in [16] is a member of the family of unstructured target detection algorithms that requires no knowledge of the background endmembers in a hyperspectral image. The CEM can be considered a filter based rather than a model based approach towards target detection.
Let $\vec{w}^{\top} = [w_1, \ldots, w_p]$ denote the coefficients of a linear finite impulse response filter $\vec{w}$. The CEM aims to estimate the filter coefficients as to maximize the energy of the filtered signal (i. e. the filtered hyperspectral observation data), when the target $\vec{t}$ is present in a pixel vector. For the filter input $\vec{x}_i \in \mathbb{R}^p$ the filter output is given by $y_i = \vec{w}^{\top}\vec{x}_i$. Hence, the average signal energy μ_E of the filtered signal amounts to

$$\begin{aligned} \mu_E &= \frac{1}{N}\sum_{i=1}^{N} y_i^2 = \frac{1}{N}\sum_{i=1}^{N} \vec{w}^{\top}\vec{x}_i\vec{x}_i^{\top}\vec{w} \\ &= \vec{w}^{\top}\Big(\frac{1}{N}\sum_{i=1}^{N} \vec{x}_i\vec{x}_i^{\top}\Big)\vec{w} = \vec{w}^{\top}\mathrm{R}_{\vec{x}}\vec{w}. \end{aligned} \tag{3.17}$$

The matrix $\mathrm{R}_{\vec{x}}$ is called the sample correlation matrix. An optimum filter should aim to minimize μ_E while transmitting the target spectrum $\vec{t}$. In mathematical terms, the latter introduces the boundary condition $\vec{w}^\top\vec{t} = 1$, and consequently:

$$\vec{w}_{\text{CEM}} = \frac{R_{\vec{x}}^{-1}\vec{t}}{\vec{t}^\top R_{\vec{x}}^{-1}\vec{t}}. \tag{3.18}$$

Thus, the CEM target detection function reads:

$$T_{\text{CEM}}(\vec{x};\vec{t}) = \vec{w}_{\text{CEM}}^\top \cdot \vec{x} = \frac{\vec{t}^\top R_{\vec{x}}^{-1}\vec{x}}{\vec{t}^\top R_{\vec{x}}^{-1}\vec{t}}. \tag{3.19}$$

3.3.5 The Orthogonal Subspace Projector

The Orthogonal Subspace Projection (OSP) algorithm was suggested by Harsanyi and Chang in [28]. Based on the linear mixture model

$$\vec{x} = \vec{t}\alpha_t + B\vec{\alpha}_B + \vec{n} \tag{3.20}$$

the algorithm belongs to the family of structured target detection algorithms capable of sub-pixel detection. Note that in contrast to the full linear mixture model that was defined for target matrices, equation 3.20 allows target vectors (as opposed to target matrices) only. The OSP algorithm is based on the idea of measuring the magnitude of a pixel vector under test, after projection onto a subspace orthogonal to the space spanned by the background vectors that form the columns of matrix B in equation 3.20.
The orthogonal complement operator, defined as

$$P_B{}^\perp = I - B(B^\top B)^{-1}B^\top \tag{3.21}$$

maps the pixel vectors of the given hyperspectral observation $\{\vec{x}_i\}_{i=1}^N$ onto a subspace orthogonal to the space spanned by the vectors that form the columns of matrix B. The target detection function of the OSP algorithm follows directly:

$$T_{\text{OSP}}(\vec{x};\vec{t}) = \frac{\vec{t}^\top P_B{}^\perp\vec{x}}{\vec{t}^\top P_B{}^\perp\vec{t}} \tag{3.22}$$

The OSP was originally proposed without the denominator $\vec{t}^{\top} P_{B}^{\perp}\vec{t}$. However, it can be shown (e. g. [42]) that under the Gaussian white noise assumption the OSP equals the maximum likelihood estimate of the target abundance $T_{OSP}(\vec{x};\vec{t}) = \hat{\alpha}_{\vec{t}}$.

3.3.6 The Adaptive Matched Subspace Detector

The AMSD belongs to the structured target detection algorithms and has the highly desirable CFAR property. Based on the LMM, the algorithm takes a likelihood ratio hypotheses testing approach. The two competing hypotheses to be discriminated for every pixel vector under test $\vec{x}_i \in \{\vec{x}_1, \dots, \vec{x}_N\}$ are

$$\begin{aligned} H_0 &: \vec{x}_i = B \cdot \vec{\alpha}_B + \vec{n} \\ H_1 &: \vec{x}_i = B \cdot \vec{\alpha}_B + T \cdot \vec{\alpha}_T + \vec{n} = S \cdot \vec{\alpha} + \vec{n}. \end{aligned} \tag{3.23}$$

Hypothesis H_0 covers the background only case, while H_1 accounts for nonzero target weights in the model, indicating presence of target spectra in the pixel vector. Note that we denote the concatenation of target and background matrix $S = [T, B]$ and the corresponding generalized abundance vector $\vec{\alpha}_S = [\vec{\alpha}_T^{\top}, \vec{\alpha}_B^{\top}]^{\top}$.

Under both hypotheses, the noise vector $\vec{n}$ is assumed to be zero mean normally distributed with $\vec{n} \sim \mathcal{N}(\vec{0}, \gamma_n^2 I)$. Both $\vec{\alpha}_B$ and $\vec{\alpha}_T$ as well as the noise variance γ_n^2 are unknown and have to be estimated for the generalized likelihood ratio test:

$$\mathrm{GLRT}(\vec{x}) = \frac{\mathcal{L}(\hat{\vec{\alpha}}, \hat{\gamma}_n^2 \,|\, \vec{x})}{\mathcal{L}(\hat{\vec{\alpha}}_B, \hat{\gamma}_{n,B}^2 \,|\, \vec{x})}. \tag{3.24}$$

The algorithm then employs the MLEs of multivariate Normal probability distribution parameters for the two hypotheses $\hat{\vec{\alpha}}_B$, $\hat{\vec{\alpha}}_T$ and $\hat{\gamma}_{\vec{n}}^2$, to determine a likelihood ratio for a specific observation.

Using the orthogonal projection operators $\mathrm{P_S} = \mathrm{I} - \mathrm{S}(\mathrm{S}^\top \mathrm{S}^{-1})\mathrm{S}^\top$ and $\mathrm{P_B} = \mathrm{I} - \mathrm{B}(\mathrm{B}^\top \mathrm{B}^{-1})\mathrm{B}^\top$ that were introduced in equation 3.21, the AMSD detection function can be rewritten as follows[43]:

$$T_{\mathrm{AMSD}}(\vec{\mathrm{x}}) = \frac{\vec{\mathrm{x}}^\top(\mathrm{P_B}^\perp - \mathrm{P_S}^\perp)\vec{\mathrm{x}}}{\vec{\mathrm{x}}^\top \mathrm{P_S}^\perp \vec{\mathrm{x}}} = [\mathrm{GLRT}(\vec{\mathrm{x}})]^{2/p} - 1. \tag{3.25}$$

It is convenient to use equation 3.25 instead of the pure GLRT equation 3.24 as the numerator and denominator terms are independent, which strongly facilitates determination of the test statistics [43].

3.3.7 The Adaptive Coherence Estimator

The Adaptive Coherence / Cosine Estimator (ACE) is a powerful CFAR target detection algorithm from the family of unstructured algorithms that was proposed by Scharf et al. in [54]. Like the AMSD detector the algorithm attempts to discriminate two hypotheses:

$$\begin{aligned} H_0 &: \vec{\mathrm{x}}_i = \vec{\mathrm{n}} \\ H_1 &: \vec{\mathrm{x}}_i = \mathrm{T} \cdot \vec{\alpha}_t + \beta\vec{\mathrm{n}}, \beta > 0 \end{aligned} \tag{3.26}$$

In equation 3.26 the interfering background clutter is modeled by the noise vector $\vec{\mathrm{n}}$ and – assuming zero mean observation data that can be obtained without loss of generality by removing the background mean $\vec{\mu}_\mathrm{B}$ from the target subspace and measurement data – the underlying distribution is assumed to be Gaussian normal, with

$$\vec{\mathrm{n}} \sim \mathcal{N}(\vec{0}, \Gamma_\mathrm{B})$$

under H_0 and

$$\beta\vec{\mathrm{n}} \sim \mathcal{N}(\vec{0}, \beta^2\Gamma_\mathrm{B})$$

under H_1.

The generalized likelihood ratio test for these hypotheses under the given constraints leads to the following target detection function for detection of the target subspace T in the hyperspectral observation vector $\vec{\mathrm{x}}$ [41]:

$$T_{\mathrm{ACE}}(\vec{\mathrm{x}},\mathrm{T}) = \frac{\vec{\mathrm{x}}^\top {\Gamma_\mathrm{B}}^{-1}\mathrm{T}(\mathrm{T}^\top {\Gamma_\mathrm{B}}^{-1}\mathrm{T})^{-1}\mathrm{T}^\top {\Gamma_\mathrm{B}}^{-1}\vec{\mathrm{x}}}{\vec{\mathrm{x}}^\top {\Gamma_\mathrm{B}}^{-1}\vec{\mathrm{x}}}. \tag{3.27}$$

If the target subspace has only one dimension $\mathrm{T} \in \mathbb{R}^{p\times 1}$, equation 3.27 can be re-written more compactly

$$T_{\mathrm{ACE}}(\vec{\mathrm{x}},\vec{\mathrm{t}}) = \frac{(\vec{\mathrm{t}}^\top {\Gamma_\mathrm{B}}^{-1}\vec{\mathrm{x}})^2}{(\vec{\mathrm{t}}^\top {\Gamma_\mathrm{B}}^{-1}\vec{\mathrm{t}})(\vec{\mathrm{x}}^\top {\Gamma_\mathrm{B}}^{-1}\vec{\mathrm{x}})}. \tag{3.28}$$

The background distribution parameters $\vec{\mu}_\mathrm{B}$ and Γ_B are in general unknown. Thus the corresponding maximum likelihood estimates $\tilde{\vec{\mu}}_\mathrm{B}$ and $\tilde{\Gamma}_\mathrm{B}$ are typically estimated from background training data and used for construction of the detector in equations 3.27 and 3.28.

3.4 Estimating Target Detection Performance

The main detection performance estimation and comparison tool used within this work are Receiver Operating Characteristics (ROCs). ROCs measure a binary classifier's performance in terms of its True Positive Rate (TPR) and False Alarm Rate (FAR).

Let $\mathrm{X} = \{\vec{\mathrm{x}}_i\}_{i=1}^N$ denote a hyperspectral observation and

$$\phi(\cdot;\vec{\mathrm{t}}) : \mathbb{R}^p \to \{0,1\}$$

denote the ground truth function with:

$$\phi(\vec{\mathrm{x}}_i;\vec{\mathrm{t}}) = \begin{cases} 1, \text{ if } \vec{\mathrm{x}}_i \text{ contains target } \vec{\mathrm{t}} \\ 0, \text{ otherwise.} \end{cases} \tag{3.29}$$

For a given classification function $\Delta_D(\cdot;\vec{t},\tau)$ with classification threshold value τ we can estimate the true positive rate:

$$TPR_{\Delta_D}(X;\vec{t},\tau) = \frac{1}{N}\sum_{i=1}^{N}\Delta_D(\vec{x}_i;\vec{t},\tau)\cdot\phi(\vec{x}_i;\vec{t}). \tag{3.30}$$

And the false alarm rate follows directly:

$$FAR_{\Delta_D}(X;\vec{t},\tau) = \frac{1}{N}\sum_{i=1}^{N}\Delta_D(\vec{x}_i;\vec{t},\tau)\cdot(1-\phi(\vec{x}_i;\vec{t})). \tag{3.31}$$

Thus, by variation of τ we obtain the value pairs $(TPR_{\Delta_D}(X;\vec{t},\tau),\ FAR_{\Delta_D}(X;\vec{t},\tau))$ that form the ROC for classifier Δ_D upon detection of target $\vec{t}$ in observation X.

3.4.1 Artificial Hyperspectral Images

In many real-world image processing applications it can be difficult or even impossible to obtain the ground truth mapping function $\phi(\cdot;\vec{t})$ in equation 3.29, i. e. there is usually no *labeled* image data available. This is also true for the task at hand – image labeling even under laboratory conditions is a highly non-trivial task. As target material can not be fixed on the substrate without fundamentally affecting the spectroscopic characteristics, shocks and vibrations during sample transport are expected to change the ground truth, even if an accurate target mapping could be generated in preparation. However, for accurate performance analysis and comparison of hyperspectral target detection methods, having labeled data is essential.

To elude this issue, it is convenient to work with artificially generated data for performance analysis. This is an especially inviting approach as the LMM provides us with a physically meaningful model for data generation.

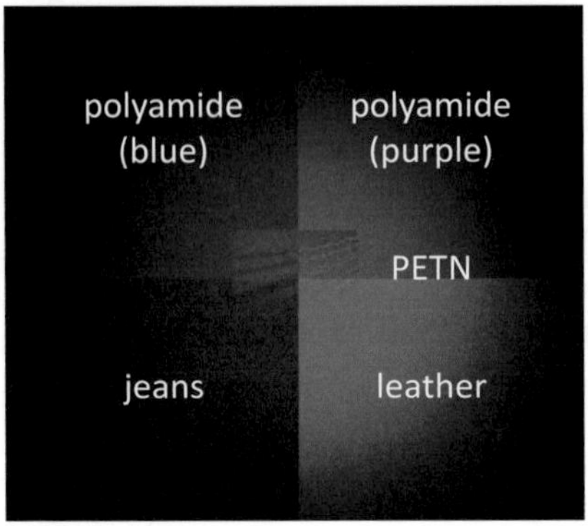

Figure 3.2: Pseudo color representation of artificial hyperspectral image containing four spatially distinct substrate material regions and a PETN contamination located in the image center in the form of the Fraunhofer logo, with decreasing abundance from top to bottom. A Gaussian illumination profile was employed for generating the shown artificial image. The pseudo colors were obtained by evenly binning the infrared channels into the channels of an RGB image.

A well defined hyperspectral image $\{\vec{x}_i\}_{i=1}^{N}$ can be generated using a slightly modified version of the LMM

$$\vec{x}_i = \Xi_i \cdot (\mathrm{T} \cdot \vec{t}_i + \mathrm{B} \cdot \vec{b}_i) + \vec{n}_i \tag{3.32}$$

by choosing the target and background abundance vectors $\vec{t}_i$ and $\vec{b}_i$ which in consequence determines the ground truth function $\phi(\cdot;\vec{t})$. The noise vectors $\vec{n}_i$ are usually generated using a Gaussian distribution. However, other distributions can be employed to assess the effects of noise model mismatches. Multiplicative noise contributions that are compensated in real-world measurement in the pre-processing step (see section 6.3) can be readily simulated by the (typically diagonal) noise matrix $\Xi_i = \text{diag}\{\vec{\xi}_i\}$ with appropriate choice of elements $\vec{\xi}_i \in \mathbb{R}^{P}_{\geq 0}$. Defective pixels are modeled by multiplicative salt and pepper noise.

A false-color representation of a typical artificial hyperspectral image generated using real substance spectra of PETN and some background materials

(Polyamide, jeans cloth and leather) is shown in figure 3.2. The PETN trace is deposited in the form of the Fraunhofer logo with decreasing abundance from top to bottom. The false colors were obtained by equidistant binning of adjacent infrared bands into RGB channels for each pixel.

3.5 Background Endmember Estimation for Structured Target Detection Algorithms

Structured target detection algorithms are reliant upon knowledge of the background spectra that occur in the hyperspectral image, as they are usually based on the full linear mixture model

$$\vec{\mathrm{x}} = \mathrm{T}\vec{\alpha}_{\mathrm{T}} + \mathrm{B}\vec{\alpha}_{\mathrm{B}} + \vec{\mathrm{n}}.$$

The matrix T denotes the notion of a target spectral library containing the pure substance spectra as columns. Likewise does B for the pure background spectra that are present in the scene. It is clear that in contrast to the target spectral library it is not a feasible approach to build a background spectral library containing all possible *background* spectra that might occur in a generic measurement scenario. Instead, B has to be generated from the hyperspectral image data $\{\vec{\mathrm{x}}_i\}_{i=1}^{N}$ to be analyzed. Algorithms that solve this task are called *endmember extractors*.

3.5.1 Using the Covariance Matrix Eigenvectors

A common way of generating the background endmember matrix B that is adopted in many publications related to structure target detection algorithms is based on the eigenvectors of the sample covariance matrix [16, 42, 57]. In this approach, the background endmember matrix is constructed using the eigenvectors that belong to the largest eigenvalues of the covariance matrix. Let $\hat{\Gamma}_{\mathrm{X}}$ denote the sample covariance matrix of a hyperspectral observation $\{\vec{\mathrm{x}}_i\}_{i=1}^{N}$ that is obtained by:

$$\hat{\Gamma}_{\mathrm{X}} = \frac{1}{N-1}\sum_{i=1}^{N}\left[(\vec{\mathrm{x}}_i - \hat{\vec{\mu}}_{\mathrm{X}}) \cdot (\vec{\mathrm{x}}_i - \hat{\vec{\mu}}_{\mathrm{X}})^{\top}\right], \tag{3.33}$$

$$\text{with } \hat{\vec{\mu}}_{\mathrm{X}} = \frac{1}{N}\sum_{i=1}^{N}\vec{\mathrm{x}}_i. \tag{3.34}$$

We denote the eigenvectors of the sample covariance matrix $\hat{\vec{v}}_1, \ldots, \hat{\vec{v}}_p$ in descending order of their corresponding eigenvalues $\hat{\lambda}_1 \geq \ldots \geq \hat{\lambda}_p$.
A reasonable choice of background matrix B is given by concatenation of the covariance matrix eigenvectors, associated to the q largest eigenvalues:

$$\mathrm{B} = [\hat{\vec{v}}_1 \cdots \hat{\vec{v}}_q]. \tag{3.35}$$

The choice of the parameter q in 3.35 is related to the material complexity of the hyperspectral image – a homogeneous background can be described by a subspace of few eigenvectors, whereas a scene containing several different background materials might require more eigenvectors. This notion is sometimes referred to as the observation's *intrinsic* or *virtual* dimensionality and will be subject to a more detailed analysis in chapter 4.

3.5.2 The Adaptive Background Generation Process

The Adaptive Background Generation Process (ABGP) [31] is an iterative and transparent background endmember extraction process that was developed within the scope of this work. The algorithm generates the background endmember matrix in two stages. In the *seeding* stage, the ABGP searches for image spectra that are most different from among each other and the target spectrum using an approach closely based on the Adaptive Target Generation Process (ATGP) proposed by Ren and Chang [51]. In the subsequent *stabilization stage* a clustering method is employed to obtain more significant class representatives.

The ATGP algorithm

Ren and Chang's ATGP algorithm is an iterative endmember extraction method that employs ideas lent of the OSP target detection approach introduced in section 3.3.5. In [51], the ATGP was suggested as first part of a fully automatic target detection process that generates a set of q potential target substance estimates $\tilde{\mathrm{T}} = [\vec{\tilde{\mathrm{t}}}_1, \ldots, \vec{\tilde{\mathrm{t}}}_q]$ within a hyperspectral image. Ren and Chang then suggest to apply a target detection algorithm, to detect the targets $\vec{\tilde{\mathrm{t}}}_i$ in a subsequent step.

Listing 1 : Original ATGP finds potential set of target spectra

Input : Hyperspectral observation $\mathrm{X} = \{\vec{\mathrm{x}}_i\}_{i=1}^N$
Output : Potential target substances $\tilde{\mathrm{T}} = [\vec{\tilde{\mathrm{t}}}_1, \ldots, \vec{\tilde{\mathrm{t}}}_q]$

$\vec{\tilde{\mathrm{t}}}_1 \leftarrow \arg\max_{\vec{\mathrm{x}}_i} \{\|\vec{\mathrm{x}}_i\|\}$ `// choose brightest pixel as first endmember`
$\tilde{\mathrm{T}} \leftarrow \vec{\tilde{\mathrm{t}}}_1$ `// initialize result matrix`
do
  `// calculate orthogonal subspace projector`
  $\mathrm{P}_{\tilde{\mathrm{T}}}{}^{\perp} \leftarrow \mathrm{I} - \tilde{\mathrm{T}}(\tilde{\mathrm{T}}^{\top}\tilde{\mathrm{T}})^{-1}\tilde{\mathrm{T}}$
  `// add vector with maximum magnitude in orthogonal subspace to result`
  $\vec{\tilde{\mathrm{t}}} \leftarrow \arg\max_{\vec{\mathrm{x}}_i} \{(\mathrm{P}_{\tilde{\mathrm{T}}}{}^{\perp}\vec{\mathrm{x}}_i)^{\top}(\mathrm{P}_{\tilde{\mathrm{T}}}{}^{\perp}\vec{\mathrm{x}}_i)\}$
  $\tilde{\mathrm{T}} \leftarrow$ `concatenate(`$\tilde{\mathrm{T}}, \vec{\tilde{\mathrm{t}}}$`)`
while $(\vec{\tilde{\mathrm{t}}}_1^{\top}\mathrm{P}_{\tilde{\mathrm{T}}}{}^{\perp}\vec{\tilde{\mathrm{t}}}_1 < \eta)$
return $\tilde{\mathrm{T}}$

The ATGP algorithm is given in listing 1. In a first step the matrix of potential target endmembers is initialized with the hyperspectral image vector containing the most signal energy. The algorithm then enters a loop, in which the orthogonal projection operator of the current target endmember matrix is calculated. The pixel vector that has maximum magnitude in the projected subspace – i. e. in the subspace orthogonal to the space spanned by the set of target spectra generated so far – is chosen and added to the potential target set. The term $\vec{\tilde{t}}_0^\top P_{\tilde{T}}^{\perp}\vec{\tilde{t}}_0$ in the cutoff criterion is a measure of similarity among the target vectors collected so far in the process that can be shown to comprise a monotonically decreasing series over the iterations.

The ABGP seeding stage

We obtain the *seeding stage* of the ABGP algorithm by modification of the ATGP. The latter can be forced to operate as a background extraction algorithm by adequate selection of some initial values. The goal of extracting background spectra of a scene implies knowledge of a set of target spectra that form the target matrix T that is considered foreground. We obtain a set of potential background material representative spectra forming the columns of the matrix $\tilde{B}$ – the *background endmember seed* – by initializing the target estimate matrix $\tilde{T}$ in listing 1 using the target matrix $T \in \mathbb{R}^{p\times t}$. Listing 2 outlines the seeding stage of the ABGP algorithm.

Listing 2 : The ABGP seeding stage as variant of the ATGP

Input : Hyperspectral observation $X = \{\vec{x}_i\}_{i=1}^N, \vec{x} \in \mathbb{R}^p$, target matrix $T \in \mathbb{R}^{p\times t}$, background dimension q

Output : Background estimate $\tilde{B} = [\vec{\tilde{b}}_1, \ldots, \vec{\tilde{b}}_q]$

`// initialize background matrix with target spectra`

$\tilde{B} \leftarrow T$

for $j \leftarrow 1$ **to** q **do**

  `// calculate orthogonal subspace projector`

  $P_{\tilde{B}}^{\perp} \leftarrow I - \tilde{B}(\tilde{B}^\top\tilde{B})^{-1}\tilde{B}$

  `// find spectrum that differs most from target and background spectra`

  $\vec{\tilde{b}} \leftarrow \arg\max_{\vec{x}_i} \{(P_{\tilde{B}}^{\perp}\vec{x}_i)^\top (P_{\tilde{B}}^{\perp}\vec{x}_i)\}$

  $\tilde{B} \leftarrow$ `concatenate`$(\tilde{B}, \vec{\tilde{b}})$

`// remove target spectra from result`

$\tilde{B} \leftarrow [\vec{\tilde{b}}_{t+1}, \ldots, \vec{\tilde{b}}_{t+q}]$

return $\tilde{B}$

We note that the ATGP cutoff criterion suggested in [51] was dropped in favor of the input parameter q that specifies the dimension of the background subspace and corresponds to the notion of *virtual dimension* introduced earlier. This was motivated by the empirical observation that in real world applications a fixed value of η showed to generalize poorly.

The ABGP stabilization stage

The result of the seeding stage is a set of background endmember spectra $\tilde{\mathrm{B}} = [\tilde{\vec{\mathrm{b}}}_1, \ldots, \tilde{\vec{\mathrm{b}}}_q]$ that could already be used as background matrix in the linear mixture model for the target detection process in this form. This is however a sub-optimal approach, as only a fraction of the available data is employed for model construction. In addition, single pixel spectra can be subject to various noise artifacts and a method like the ATGP that is based on an extreme value measure intrinsically favors outliers. This can substantially affect the quality of $\tilde{\mathrm{B}}$ as background representation.

We therefore propose to employ a stabilization stage that adopts a clustering algorithm on the observation data using $\tilde{\mathrm{B}}$ as seed vectors. Rather than the seeding vectors, we then choose the cluster means as background endmembers.

A seed-based strict partitioning of the observation data $\{\vec{\mathrm{x}}_i\}_{i=1}^{N}$ into $q+t$ (t denoting the number of columns of the target matrix T) clusters $\{C_1 \ldots C_q, C_{q+1}, \ldots, C_{q+t}\}$ can be obtained using any of the full-pixel target detection algorithms presented in section 3.3. Let $T_D(\vec{\mathrm{x}}_i; \vec{\mathrm{v}})$ denote the corresponding target detection function for the observation vector $\vec{\mathrm{x}}_i$ and target vector $\vec{\mathrm{v}}$. As the value of T_D relates to the probability that $\vec{\mathrm{x}}_i$ contains $\vec{\mathrm{v}}$ it can be employed to generate meaningful clusters.

Based on the cumulative set of seed vectors and target spectra $\mathrm{V} = \{\tilde{\vec{\mathrm{b}}}_1, \ldots, \tilde{\vec{\mathrm{b}}}_q, \vec{\mathrm{t}}_1, \ldots, \vec{\mathrm{t}}_t\}$ we gain the strict partitioning

$$C_k = \left\{ \vec{\mathrm{x}}_i \;\middle|\; k = \underset{k \in \{1,\ldots,q+t\}}{\arg\max}\; T_D\left(\vec{\mathrm{x}}_i; \mathrm{V}[k]\right) \right\} \tag{3.36}$$

The target clusters $C_{q+1}, \ldots, C_{q+t}$ are considered to contain one of the target spectra and are hence collected into a rejection class, having no further impact on the result. The columns of the endmember matrix $\mathrm{B} = [\vec{\mathrm{b}}_1, \ldots, \vec{\mathrm{b}}_q]$ are finally obtained by the remaining cluster means:

$$\vec{\mathrm{b}}_k = \frac{1}{|C_k|} \sum_{\vec{\mathrm{x}}_l \in C_k} \vec{\mathrm{x}}_l. \tag{3.37}$$

As stated before, all full-pixel target detection functions can be employed for the choice of T_D in equation 3.36. For computation performance reasons, we chose the most straightforward approach to full-pixel target detection, comprised by the NCC algorithm.

3.6 Target Leakage Prevention

The main difference between *structured* and *unstructured* target detection algorithms is the modeling of the background clutter. Mathematically this leads to the definition of the *full* and *reduced* linear mixture model. The former is adopted by the structured target detection algorithms and explicitly requires knowledge of the matrix B of background spectra. In contrast, the latter assumes that the target signal is embedded in random noise that is usually assumed to be Gaussian Normal distributed. However, the corresponding noise parameters (background mean vector and covariance matrix) are in general unknown. Ideally, a background training set would be available for every hyperspectral image, containing it's background spectra only, from which the parameters can be estimated. However, in many applications this is not a feasible demand. Instead, the background distribution parameters have to be estimated from the hyperspectral observation under analysis. This problem is obviously closely related to background endmember extraction which was introduced in section 3.5 though now put into more statistical rather than geometrical terms.

We say that *target sparsity* is given, if a target material is rare in comparison to other spectra within a hyperspectral image. In mathematical terms, the target $\vec{\mathrm{t}}$ is said to be sparse in $\mathrm{X} = \{\vec{\mathrm{x}}_i\}_{i=1}^N$, if the mean magnitude of the background abundance vector is significantly higher than the target abundance mean. Thus, if the observation X is fully described by the linear mixture model

$$\vec{\mathrm{x}}_i = \vec{\mathrm{t}} \cdot t_i + \mathrm{B} \cdot \vec{\mathrm{b}}_i + \vec{\mathrm{n}} \qquad (3.38)$$

where t denotes the target abundance and $\vec{\mathrm{b}}_i$ the background abundance vector, target sparsity is given if

$$\frac{1}{N}\sum_{i=1}^{N} t_i << \frac{1}{N}\sum_{i=1}^{N} ||\vec{\mathrm{b}}_i||. \qquad (3.39)$$

If target sparsity can be assumed, the parameters for the background noise distribution can be safely estimated from the entire observation X. If however, the target is not sparse, the noise distribution estimates are corrupted and target detection performance will be strongly affected. Hence, the goal within this section is to suggest – given a target spectrum $\vec{\mathrm{t}}$ – a binary *background decision function* $\beta(\cdot;\vec{\mathrm{t}})$ that serves to discriminate a hyperspectral observation $\mathrm{X} = \{\vec{\mathrm{x}}_i\}_{i=1}^N$ into a background set $\mathrm{X}_{\mathrm{B},\vec{\mathrm{t}}}$ and a set $\mathrm{X}_{\mathrm{T},\vec{\mathrm{t}}}$ containing potentially contaminated spectra:

$$\mathrm{X}_{\mathrm{B},\vec{\mathrm{t}}} := \{\vec{\mathrm{x}} \in \mathrm{X} \mid \beta(\vec{\mathrm{x}};\vec{\mathrm{t}}) = 1\} \text{ and } \mathrm{X}_{\mathrm{T},\vec{\mathrm{t}}} := \{\vec{\mathrm{x}} \in \mathrm{X} \mid \beta(\vec{\mathrm{x}};\vec{\mathrm{t}}) = 0\} \qquad (3.40)$$

$$\text{with } \beta(\vec{\mathrm{x}};\vec{\mathrm{t}}) = \begin{cases} 1 & \vec{\mathrm{x}} \text{ not contaminated by } \vec{\mathrm{t}} \\ 0 & \text{otherwise.} \end{cases} \qquad (3.41)$$

The background noise distribution parameters can then be estimated from the background set $X_{B,\vec{t}}$. The quality of the background set and subsequently the quality of the distribution parameter estimates is determined by the background detection function, which thus directly affects the expected target detection performance. The term *target leakage* refers to an incorrect mapping of the observation data, where contaminated pixel vectors are falsely classified as background spectra, thus compromising the background distribution parameter estimates. We call $\beta(\cdot;\cdot)$ target leakage prevention function.

The definition of the background decision function $\beta(\cdot;\cdot)$ is similar to the general target classification function $\tau(\cdot;\cdot)$ defined in equation 3.9. At a first glance, one might be inclined to believe that the principle of target detection is basically the negation of background detection in which case the task of target leakage prevention would induce a logical circle. This is not the case for two reasons. Firstly, the boundary conditions for the two tasks are slightly different. Secondly, background detection is only relevant for algorithms from the family of unstructured methods that are capable of sub-pixel target detection. Of course, comparing the expected true positive detection rate, in presence of sub-pixel contaminations, such algorithms are expected to outperform full-pixel detectors. The latter however, do not require knowledge of the background at all and are thus potential candidates for background detection.

The expected true-positive and false alarm rate of classification functions are directly dependent upon the threshold τ (refer to section 3.4 for more details). Reducing the threshold will in general increase the TPR as well as the FAR. Using the negated output of a target detection function at a low threshold as background classification function is a feasible way of strongly reducing target leakage. False alarms of the detection function should be to a certain extent acceptable in the background classification function, their only effect being that some of the background spectra are not used for estimation of the distribution parameters.

We therefore suggest to adopt the negative output of the Normalized Cross Correlation function as target leakage prevention function:

$$\beta_{\mathrm{NCC}}(\vec{\mathrm{x}};\vec{\mathrm{t}}) = -1 \cdot T_{\mathrm{NCC}}(\vec{\mathrm{x}};\vec{\mathrm{t}}). \tag{3.42}$$

Other choices could include the inverse of the CEM detection function as well as variants of the MF [58].

3.7 Detector Fusion

As mentioned before, the task of target detection and target identification intrinsically carries the notion of a target library. I. e. a hyperspectral observation $\mathrm{x} = \{\vec{\mathrm{x}}_i\}_{i=1}^{N}$ should be tested for contamination with one or many of a set of possible target materials, for which the spectra are known. Let $\mathrm{T} = \left[\vec{\mathrm{t}}_1, \cdots, \vec{\mathrm{t}}_t\right]$ denote this target spectral library containing the spectra of t substances of interest. However, most target detection algorithms allow only a single target spectrum for operation. Hence, a detection output is generated for each pixel vector in the set and each target in the library.
We recall the generic target detection function given in section 3.3 that can be written as

$$T_D : \quad (\vec{\mathrm{x}}_i \in \mathrm{X};\vec{\mathrm{t}}_k) \quad \rightarrow \quad \mathbb{R} \tag{3.43}$$

Equation 3.43 states that a detection function produces a scalar value for every pixel vector and target spectrum.
Hence, for the case of a target *library* T we extend this function to

$$\vec{\mathrm{T}}_D : \quad (\vec{\mathrm{x}}_i \in \mathrm{X};\mathrm{T}) \quad \rightarrow \quad \mathbb{R}^t \tag{3.44}$$

Consequently, for every pixel vector $\vec{\mathrm{x}}_i \in \mathrm{X}$ we obtain a detection result vector of the form

$$\vec{\mathrm{T}}_D(\vec{\mathrm{x}}_i \in \mathrm{X};\mathrm{T}) = \left[\, T_D(\vec{\mathrm{x}}_i;\vec{\mathrm{t}}_1), \cdots, T_D(\vec{\mathrm{x}}_i;\vec{\mathrm{t}}_t) \,\right]^{\top} \tag{3.45}$$

We obtain the corresponding library *classification function* by choosing the index of the target spectrum $\vec{\mathrm{t}}_k$, with the largest detection function output – if it is larger than a specific threshold τ_k. In mathematical terms:

$$\Delta_D(\vec{\mathrm{x}}_i; \mathrm{T}, \vec{\tau} \in \mathbb{R}^t) := \begin{cases} \arg\max_k D_T(\vec{\mathrm{x}}; \vec{\mathrm{t}}_k), & \text{if } D_T(\vec{\mathrm{x}}; \vec{\mathrm{t}}_k) > \tau_k \\ 0, & \text{else.} \end{cases} \tag{3.46}$$

Hence, the classification function is no longer binary, but a mapping into the set $\{0, 1, \ldots, t\}$ that contains the indices of the corresponding target spectra. The zero index indicates the background class.

The threshold values $\tau_1, \ldots, \tau_t$ can be chosen individually, i. e. selecting a specific threshold for each target spectrum. This approach may be beneficial in highly controlled application scenarios as a-priori knowledge can be implemented by careful selection – i. e. higher false alarm rates (and thus lower thresholds) or vice versa may be acceptable for some substances in comparison to others. The generic measurement scenario that we are aiming for within this work however, suggests to choose $\tau_1 = \tau_2 = \ldots = \tau_t = \tau$.

4 The Problem of Model Order Selection

In section 3.5 we closed with the remark that the only remaining model parameter to be estimated is the column dimension of the background endmember matrix, i. e. the parameter q in $\mathrm{B} \in \mathbb{R}^{p \times q}$ in the full mixture model. Hence, we are searching for a rule to determine the parameter q given the linear model

$$\vec{\mathrm{x}}_i = \mathrm{S} \cdot \vec{\alpha}_i + \vec{\mathrm{n}}_i,\ \mathrm{S} \in \mathbb{R}^{p \times q},\ \vec{\alpha}_i \in \mathbb{R}^q, \vec{\mathrm{n}} \sim \mathcal{N}(\vec{0}, \sigma^2 \mathrm{I}). \tag{4.1}$$

This corresponds to the notion of *intrinsic dimension* or *virtual dimensionality* that has been subject to some research in the past (e. g. [3, 4, 7, 11, 12]). The column order of the material matrix S corresponds to the number of spectrally distinct materials in the scene and determines the model order. Hence, determining q can be considered to comprise a model order selection problem.

In this section we will investigate an approach based on the Minimum Description Length (MDL) principle that was briefly considered for the task of model order selection or *virtual dimension estimation* by Broadwater in [7]. Broadwater's work is a modification of an approach suggested by Wax and Kailath [63] who tackled a similar problem of estimating the number of independent signal sources in complex radar signals. Broadwater eventually rejected the method as dysfunctional. However, this can be contributed to an error that occurred by modifying Wax' formulation for complex signals to real valued observations.

4.1 Considerations on the Eigenvalue Distribution of Covariance Matrices

Most approaches to linear model order selection essentially analyze the eigenvalue distribution of the sample covariance matrix, as the parameter q is directly encoded into it. To see this, we follow Wax' and Kailath's derivation given in [63] and consider the real-valued multi-band observation $\mathrm{X} = \{\vec{\mathrm{x}}_i\}_{i=1}^{N}, \vec{\mathrm{x}}_i \in \mathbb{R}^p$ in the light of the linear model in its most general form given in 4.1. Assuming a zero-mean noise distribution with diagonal covariance matrix and constant variance, we obtain a model for the covariance matrix $\mathrm{R} \in \mathbb{R}^{p \times p}$ of X as

$$\mathrm{R} = \Psi + \sigma^2 \cdot \mathrm{I}, \text{ with } \Psi = \mathrm{S}\mathrm{C}_{\vec{\alpha}}\mathrm{S}^\top. \tag{4.2}$$

In equation 4.2 $\mathrm{C}_{\vec{\alpha}}$ denotes the covariance matrix of the abundance vectors

$$\mathrm{C}_{\vec{\alpha}} = \mathscr{E}\{(\vec{\alpha}_i - \vec{\mu}_\alpha) \cdot (\vec{\alpha}_i - \vec{\mu}_\alpha)^\top\}, \tag{4.3}$$

where $\vec{\mu}_\alpha$ denotes the abundance mean.

We assume that the columns comprising S are linearly independent and hence, S being of full column rank. Also, we assume, that the abundance covariance matrix $\mathrm{C}_{\vec{\alpha}}$ is non-singular. It follows that Ψ has rank q and the $p - q$ smallest eigenvalules of Ψ equal zero. Let $\{\lambda_1, \ldots, \lambda_p\}$ denote the eigenvalues of R with $\lambda_1 \geq \cdots \geq \lambda_p$. Then

$$\lambda_{q+1} = \cdots = \lambda_p = \sigma^2. \tag{4.4}$$

To conclude, equation 4.4 states that in a multi-band observation following the linear model with Gaussian random noise distribution $\vec{\mathrm{n}} \sim \mathscr{N}(\vec{0}, \sigma^2\mathrm{I})$, the eigenvalues corresponding to the noise eigenvectors are all equal to the noise variance σ^2.

We note that equation 4.2 serves as generic model for a covariance matrix whereas we indicate the sample covariance matrix of observation $\mathrm{X} = \{\vec{\mathrm{x}}_i\}_{i=1}^{N}$ by a hat:

$$\hat{\mathrm{R}} = \frac{1}{N-1} \sum_{i=1}^{N} \left(\vec{\mathrm{x}}_i - \hat{\vec{\mu}}_{\mathrm{X}}\right) \cdot \left(\vec{\mathrm{x}}_i - \hat{\vec{\mu}}_{\mathrm{X}}\right)^{\mathsf{T}} . \tag{4.5}$$

Likewise we write $\{\hat{\lambda}_1, \ldots, \hat{\lambda}_p\}$ to denote the set of sample covariance matrix eigenvalues.

4.2 Existing Model Order Selection Methods

In this section we introduce three well-known approaches to model order selection in hyperspectral images, that are based on an analysis of the sample covariance matrix eigenvalue distribution. We then present a derivation of an approach developed within this work, which is based upon the MDL principle. We note, that whereas all of the following approaches require freely chooseable design parameters, the proposed MDL based approach is entirely parameter free.

4.2.1 Principal Components Energy Metric

The Principal Components Analysis (PCA) energy metric approach follows directly from the observations on the covariance matrix eigenvalue distribution given in the introduction. Under the assumption that the noise eigenvalues are small in comparison to the signal eigenvalues it is suggested to apply a threshold to the *total eigenvalue energy ratio*, which is defined as the fraction of the cumulative sum of the ordered eigenvalues to the total eigenvalue sum. The cutoff index serves as model order estimate.

In mathematical terms, let $\{\hat{\lambda}_i\}_{i=1}^{p}$ denote the ordered set of sample covariance matrix eigenvalues $\hat{\mathrm{R}}$ with $\hat{\lambda}_i \geq \hat{\lambda}_{i+1}$. Using the total eigenvalue energy ratio

$$\rho_k = \frac{\sum_{i=1}^{k} \hat{\lambda}_i}{\sum_{i=1}^{p} \hat{\lambda}_i}, \tag{4.6}$$

the model order estimate $\tilde{q}_{PCA}$ is then defined as

$$\tilde{q}_{PCA} = \arg\min_{k} [\rho_k \geq \tau]. \tag{4.7}$$

Typical choices of the threshold value are $\tau = 0.99$ or $\tau = 0.999$. We note that the result $\tilde{q}_{PCA}$ depends on the signal to noise ratio of the measurement X. Figure 4.1 illustrates this fact in a simple simulation. Two artificial hyperspectral observations with $q = 5$ endmembers each, were generated (see section 3.4.1 for more details on artificial image data). The Signal to Noise Ratio (SNR) was set to 30 dB and 10 dB respectively. We give the resulting sample covariance matrix eigenvalue distributions in figure 4.1 as well as the energy ratio that was considered as measure of model order in this section. Whereas the general form of the eigenvalue distribution is qualitatively very similar for the two noise levels, the noise eigenvalues carry more energy in the latter case. Hence, different thresholds would be required to estimate the correct model order $q = 5$ for the two observations.

4.2.2 Noise Subspace Projection

The Noise Subspace Projection (NSP) approach to model order selection in hyperspectral images is based on statistical hypothesis testing published by Chang and Du in [10]. It builds upon the earlier proposed Harsanyi-Farrand-Chang (HFC) method that was suggested by Harsanyi et al. in [27].

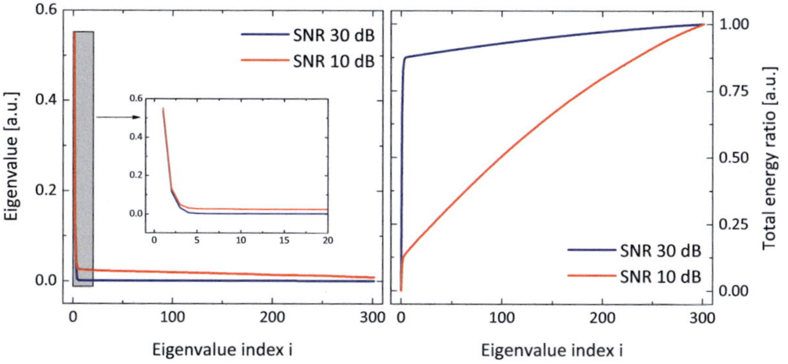

Figure 4.1: Impact of different levels of additive white Gaussian noise on the sample covariance matrix eigenvalue distribution. Left: general form of eigenvalue distribution is mostly independent, but more energy lies in eigenvalues corresponding to noise eigenvectors. Right: the total energy ratio to discriminate noise eigenvalues from signal eigenvalues is strongly dependent upon noise level. The eigenvalues were obtained from an artificial hyperspectral image generated with $q = 5$ endmembers.

The HFC compares the eigenvalues generated by the sample covariance matrix $\hat{\mathrm{R}}$ denoted $\hat{\lambda}_1, \cdots, \hat{\lambda}_p$ to the eigenvalues $\hat{\gamma}_1, \cdots, \hat{\gamma}_p$ of a matrix $\hat{\mathrm{C}} \in \mathbb{R}_{\geq 0}^{p \times p}$ that is defined as follows:

$$\hat{\mathrm{C}} := \frac{1}{N} \sum_{i=1}^{N} \vec{\mathrm{x}}_i \vec{\mathrm{x}}_i^{\top}. \tag{4.8}$$

The matrix $\hat{\mathrm{C}}$ is called *correlation matrix* in several publications (e.g. [12]), which is as pointed out in [3] however a misleading term as the entries are not correlations.

Harsanyi et al. argued that equality $\hat{\gamma}_i = \hat{\lambda}_i$ holds, if the eigenvalues correspond to noise eigenvectors. This gives rise to the following series of p hypothesis tests:

$$\begin{aligned} H_0 &: z_i = \hat{\gamma}_i - \hat{\lambda}_i = 0 \\ H_1 &: z_i = \hat{\gamma}_i - \hat{\lambda}_i > 0. \end{aligned} \tag{4.9}$$

Zero mean normal distributions $p_0(z_i) = \mathcal{N}(0, \sigma_{z_i}^2)$ and $p_1(z_i) = \mathcal{N}(\mu_i, \sigma_{z_i}^2)$ are assumed, with the parameter $\sigma_{z_i}^2$ being estimated to

$$\hat{\sigma}_{z_i}^2 = \frac{2}{N}(\hat{\gamma}_i^2 + \hat{\lambda}_i^2 + 2\hat{\gamma}_i\hat{\lambda}_i). \tag{4.10}$$

A Neyman-Pearson detector with an adaptive threshold based on the expected cumulative probability of false alarm is constructed for each eigenvalue pairing. The number of tests for which H_0 was rejected is employed as resulting model order estimate.
The NSP algorithm suggested by Chang and Du takes a similar approach to model order selection that operates on a whitened version of the sample covariance matrix. Using an estimate of the noise covariance matrix $\hat{\mathrm{R}}_n$ the whitened sample covariance matrix $\hat{\mathrm{R}}_w$ is obtained by

$$\hat{\mathrm{R}}_w = \hat{\mathrm{R}}_n^{(-1/2)} \hat{\mathrm{R}} \hat{\mathrm{R}}_n^{(-1/2)} \tag{4.11}$$

More details to the construction of the whitening matrix $\hat{\mathrm{R}}_n^{(-1/2)}$ will be given later in section 4.4.4. The noise-related eigenvalues of the whitened sample covariance matrix are expected to equal one. Hence, the competing hypotheses can be re-written to

$$\begin{aligned} H_0 &: y_i = \hat{l}_i = 1 \\ H_1 &: y_i = \hat{l}_i > 1, \end{aligned} \tag{4.12}$$

where the $\hat{l}_i$ denote the eigenvalues generated by $\hat{\mathrm{R}}_w$.

Now, $p_0(y_i) = \mathcal{N}(1, \sigma_{y_i}^2)$ and $p_1(y_i) = \mathcal{N}(\mu_i, \sigma_{y_i}^2)$, with the distribution parameter

$$\hat{\sigma}_{y_i}^2 \sim \frac{2\hat{l}_i^2}{N}. \tag{4.13}$$

Again, the cumulative probability of false alarms is employed to establish a dynamic threshold value for each test, and the number of tests for which H_0 is rejected serves as resulting model order estimate. The test statistics involved in selection of the adaptive threshold values requires choice of the marginal probability value (p-value) at which the null hypothesis is rejected. This comprises a design parameter that needs to be chosen at implementation time and strongly affects the estimate.

4.2.3 Second Moment Linear Dimension

The Second Moment Linear Dimension (SML) approach to model order selection was proposed by Bajorski in [4]. Bajorski based the SML algorithm on a purely empirical observation: a specific function of the sample covariance matrix eigenvalues follow a Gamma distribution, if they are associated to noise eigenvectors.

Let $\{\hat{\lambda}_i\}_{i=1}^{p}$ denote the ordered set of sample covariance matrix eigenvalues $\hat{\mathrm{R}}$ with $\hat{\lambda}_i \geq \hat{\lambda}_{i+1}$. The SML utilizes as measure of the relative gap size between two eigenvalues the ratio

$$f_j := (\hat{\lambda}_j - \hat{\lambda}_{j+1})/\hat{\lambda}_{j+1}. \tag{4.14}$$

According to Bajorski, empirical studies showed that the proportions f_j follow a Gamma distribution, if the participating eigenvalues belong to noise eigenvectors. Hence, he proposes to apply sequential statistical hypotheses testing on the ordered set of eigenvalues, to determine, whether the underlying distribution can be considered to be the Gamma distribution. A marginal threshold on the p-score serves as cutoff criterion for the test sequence and estimates the model order.

Listing 3 : The SML algorithm for model order selection [4]

Input : Hyperspectral observation $\mathrm{X} = \{\vec{\mathrm{x}}_i\}_{i=1}^{N}, \vec{\mathrm{x}}_i \in \mathbb{R}^p$, marginal p-value threshold τ, eigenvalues of sample covariance matrix $\{\hat{\lambda}_1, \ldots, \hat{\lambda}_p\}$
Output : Estimated model order q_{SML}
$m \leftarrow \lceil \frac{p}{3} \rceil$
$j \leftarrow p - m$; **do**
  $j \leftarrow j - 1$
  p-value $\leftarrow testGammaDistribution(\{\hat{\lambda}_j \ldots, \hat{\lambda}_p\})$
while p-value $\leq \tau$
return j

A pseudo-code implementation of the SML algorithm is given in listing 3. Details on the hypotheses test that involve distribution parameter estimation and calculation of the p-value, are omitted to improve readability and can be found in [4].

We emphasize at this point that all considered model order selection methods require selection of a parameter that has to be chosen at design-time and affects the resulting model order estimate. Especially in complex systems that involve several independent data analysis stages and operate in generic scenarios it is desirable to avoid methods that come at the cost of free parameters. The following MDL based model order selection method is not subject to this problem.

4.3 The Two-Part MDL Principle

The Minimum Description Length (MDL) principle is an information-theoretic approach to model selection that was first published by Schwartz and Rissanen in 1978 ([52, 55]) and has since been subject to various refinements and improvements ([5, 25, 26]). MDL found strong resonance in the radar signal analysis community not least to Wax' and Kailath's ground breaking work [63], who published a MDL based method for estimating the number of distinct signal sources in complex-valued radar signals.

Throughout the following chapter we will adopt Grünwald's terminology [25] that puts the terms

- *model* for a family of probability distributions or functions with the same functional form (e. g. the linear mixture model with unspecified parameter q)
- *point hypothesis* for a specific realization of a model
- *hypothesis* as generic term for either one.

Thus, by this definition we use the term *model selection problem*, instead of the slightly more cumbersome *model order selection problem*.
The basic idea behind MDL model selection is to find the best hypothesis, among a parametric model that allows to describe the combination of both the hypothesis (including all parameters) as well as the observation data with a minimum number of information theoretic bits. To support this notion we borrow an example from Grünwald's tutorial on MDL [25]: Let $\mathscr{H}^{(1)}, \mathscr{H}^{(2)}, \ldots$ denote sets of candidate models, considered to describe the observation data X and $\mathscr{H}^{(k)}$ denotes the set of k order models (e. g. the linear mixture model with $\mathrm{S} \in \mathbb{R}^{p \times k}$). We are then searching for the best hypothesis $H \in \bigcup_i \mathscr{H}^{(i)}$ that minimizes the sum $L(H) + L(\mathrm{X} \mid H)$, where

- $L(H)$ is the coding length in bits of the hypothesis and all its parameters, and
- $L(\mathrm{X} \mid H)$ is the length in bits of the hypothesis description

I. e. the MDL follows the idea that the best model to explain the observed data is the smallest model containing the selected hypothesis. It hence searches for a tradeoff between goodness-of-fit and complexity of the models involved an can be considered an implementation of Occam's razor. The function $L(\mathrm{X} \mid H)$ resembles the codelength of the observation X, when encoded with the help of the hypothesis H.

For probabilistic hypotheses – and we shall restrict ourselves to such hypotheses in the following – the optimum code is the Shannon-Fano code [15, 56] for which the codelength can be calculated explicitly:

$$L(\mathrm{X} \mid H) = -\log p(\mathrm{X} \mid H),$$

where $p(\mathrm{X} \mid H)$ denotes the probability density function to the corresponding hypothesis.
The hypothesis codelength $L(H)$ is not quite as straightforward to obtain and often amounts to counting the free parameters of hypothesis H. Rissanen suggested for an observation size N

$$L(H) = k \cdot \log \frac{N}{2\pi}$$

with k denoting the number of free parameters in H. This choice of model codelength leads to what is now generally referred to as *Two-Part MDL*:

$$MDL_{2P} = -\log p(\mathrm{X} \mid H) + k \cdot \log \frac{N}{2\pi} \tag{4.15}$$

The first term in equation 4.15 denotes the negative log-likelihood function and resembles the number of bits required to code the model error or goodness of fit. The second term estimates the number of bits required to code the model description. I. e. a very complex model may yield a good fit and thus reduce the required number of bits to encode the model error. However, as many parameters have to be described, the model codelength will increase. In this sense, the model codelength term serves as penalty function for overly complex models to intrinsically avoid over-fitting.

4.4 MDL in Linear Model Dimension Estimation

In this section we derive a model selection method based on the MDL principle, for estimating the number of endmembers in a hyperspectral observation. We note that the method is not restricted to hyperspectral images, but is applicable to any real-valued multi-band measurement observation that can be described by a linear superposition of independent signal sources. The approach closely follows Wax' and Kailath's suggestion, who published a method for estimating the number of signal sources in radar applications that are typically complex-valued. Following section 4.3 the two part MDL for probabilistic hypotheses aims to minimize the sum

$$-\log p(\mathrm{X} \mid \vec{\theta}) + \frac{1}{2}\kappa(\vec{\theta})\log N. \tag{4.16}$$

where $p(\cdot \mid \cdot)$ denotes the probability density function of choice, $\vec{\theta}$ the corresponding parameter vector, and $\kappa(\cdot)$ denotes the number of free parameters in the parameter vector.

The first part of the sum in equation 4.16 is the data codelength term and resembles the number of bits required to code the model error or goodeness of fit. As we restrict ourselves to probabilistic models, by definition it is minimized by the appropriate MLE of $\vec{\theta}$ that we denote $\hat{\vec{\theta}}$. The second term – the model codelength – estimates the number of bits required to encode the model description length including all required parameter values.

4.4.1 Data Codelength

A parametric model for the covariance matrix R given in equation 4.2 is [63]

$$\mathrm{R}^{(k)} = \Psi^{(k)} + \sigma^2\mathrm{I} \tag{4.17}$$

where the parameter k drives the rank of $\Psi^{(k)}$.

Then $\mathrm{R}^{(k)}$ can be expressed in terms of its eigenvectors $\vec{v}_1, \dots, \vec{v}_k$ with $\vec{v}_i \in \mathbb{R}^p$ according to the ordered eigenvalues $\lambda_1 \geq \dots \geq \lambda_k$:

$$\mathrm{R}^{(k)} = \sum_{i=1}^{k} (\lambda_i - \sigma^2) \vec{v}_i \vec{v}_i^\top + \sigma^2 \mathrm{I}. \tag{4.18}$$

This directly provides a reasonable choice of parameter vector:

$$\vec{\theta}^{(k)} = (\lambda_1, \dots, \lambda_k, \sigma^2, \vec{v}_1^\top, \dots, \vec{v}_k^\top)^\top. \tag{4.19}$$

We choose the family of Gaussian probability density functions as probabilistic model. Omitting most of the details of the derivation that are given in appendix A, we obtain the following explicit form of the log-likelihood function $\mathscr{L}(\cdot \mid \vec{\theta}^{(k)})$:

$$\begin{aligned} \mathscr{L}(\mathrm{X} \mid \vec{\theta}^{(k)}) &= \log p(\vec{\mathrm{x}}_1, \dots, \vec{\mathrm{x}}_N \mid \vec{\theta}^{(k)}) = \log \prod_{i=1}^{N} p(\vec{\mathrm{x}}_i \mid \vec{\theta}^{(k)}) \\ &= -\frac{N}{2} \left[\log (2\pi)^p + \log \det \mathrm{R}^{(k)} + \mathrm{tr} \left[\mathrm{R}^{(k)^{-1}} \cdot \tilde{\mathrm{R}} \right] \right] \\ &= -\frac{N}{2} \left[\log (2\pi)^p + \sum_{i=1}^{k} \left[\log \lambda_i + \frac{l_i}{\lambda_i} \right] + (p-k) \cdot \log \sigma^2 + \sum_{i=k+1}^{p} \frac{l_i}{\sigma^2} \right]. \end{aligned} \tag{4.20}$$

The MLEs that maximize the log-likelihood function read [2]

$$\begin{aligned} \hat{\lambda}_i &= l_i, i = 1, \dots, k, l_i \in \mathbb{R} \\ \hat{\sigma}^2 &= \frac{1}{p-k} \sum_{i=k+1}^{p} l_i \\ \hat{\vec{v}}_i &= \vec{\mathrm{c}}_i, i = 1, \dots, k, \vec{\mathrm{c}}_i \in \mathbb{R}^p. \end{aligned} \tag{4.21}$$

The l_i denote the eigenvalues to the sample covariance matrix eigenvectors $\vec{\mathrm{c}}_i$. Plugging the MLEs in equations 4.21 into the log-likelihood function –

and once more omitting the details that are given in appendix A – we obtain the data code length

$$\mathscr{L}(\hat{\vec{\theta}}) = -\frac{N}{2}\left[\sum_{i=1}^{k} \log \lambda_i + (p-k) \cdot \log \sum_{i=k+1}^{p} \frac{l_i}{p-k}\right] + c \tag{4.22}$$

where we have collected all terms that do not depend on k in the constant c. This can be safely dropped in the minimization step, as it has no impact on the result.

4.4.2 Model Code Length

The next task stands to determine the number of free parameters $\kappa(\vec{\theta}^{(k)})$ in the model parameter vector $\vec{\theta}^{(k)}$ as is required for the model code length term in equation 4.16. For easier readability we restate the definition of the model parameter vector

$$\vec{\theta}^{(k)} = (\lambda_1, \ldots, \lambda_k, \sigma^2, \vec{v}_1^\top, \ldots, \vec{v}_k^\top)^\top.$$

It is obvious that we require $k+1$ free parameters to account for the eigenvalues $\lambda_1, \ldots, \lambda_k$ and the noise variance σ^2. This yields

$$\kappa(\vec{\theta}^{(k)}) = k + 1 + N(\{\vec{v}_i\}_{i=1}^{k}), \tag{4.23}$$

where $N(\{\vec{v}_i\}_{i=1}^{k})$ denotes the number of free adjustable parameters of the set of eigenvectors $\{\vec{v}_i\}_{i=1}^{k}$. Being eigenvectors to different eigenvalues of $\mathrm{R}^{(k)}$ which is symmetric and positive-semidefinite, the vectors $\{\vec{v}_i\}_{i=1}^{k}$ are consequently mutually orthogonal and it hence suffices to calculate the dimension of the group of real valued $(p \times k)$ matrices with mutually orthogonal columns. More formally:

$$N(\{\vec{v}_i\}_{i=1}^{k}) = \dim\{\mathrm{M} \in \mathbb{R}^{p \times k} \,|\, \mathrm{M}^\top \mathrm{M} = \mathrm{I}\}. \tag{4.24}$$

The group on the right hand side of equation 4.24 is known as Stiefel-Manifold over the field $\mathbb{R}$:

$$\mathrm{V}_k(\mathbb{R}^p) = \{\mathrm{M} \in \mathbb{R}^{p\times k} \,|\, \mathrm{M}^\top \mathrm{M} = \mathrm{I}\} \tag{4.25}$$

with the known dimension

$$\mathrm{V}_k(\mathbb{R}^p) = pk - \frac{1}{2}k(k+1). \tag{4.26}$$

Using these results, equation 4.23 simplifies to

$$\kappa(\vec{\theta}^{(k)}) = k + 1 + pk - \frac{1}{2}k(k+1). \tag{4.27}$$

We now plug the result from equations 4.22 and 4.27 into the MDL equation 4.16 and (omitting the terms that do not depend on k) obtain:

$$\begin{aligned} MDL_{2P} &= -\mathscr{L}(\mathrm{X} \,|\, \hat{\vec{\theta}}^{(k)}) \\ &= -\frac{N}{2}\left(\sum_{i=1}^{k} \log l_i + \log\left[\sum_{i=k+1}^{p} \frac{l_i}{p-k}\right]^{p-k}\right) \\ &\quad + \frac{1}{2}(k + pk - \frac{1}{2}k(k+1)) \log N. \end{aligned} \tag{4.28}$$

Apart from the sample covariance matrix eigenvectors and -values, equation 4.28 depends on the parameter $k \in \{1, \dots, p\}$ only. Following the MDL principle we calculate the model order and i. e. the number of columns of the material matrix S in the linear mixture model as

$$\begin{aligned} q_{MDL} = \arg\min_k \Bigg[&-\frac{N}{2}\left(\sum_{i=1}^{k} \log l_i + \log\left[\sum_{i=k+1}^{p} \frac{l_i}{p-k}\right]^{p-k}\right) \\ &+ \frac{1}{2}(k + pk - \frac{1}{2}k(k+1)) \log N \Bigg]. \end{aligned} \tag{4.29}$$

4.4.3 Comparison of Model Codelength Terms

In the previous section we diverted considerably from Wax' and Kailath's estimation of free parameters κ in the model parameter vector $\vec{\theta}^{(k)}$. In [63] the authors estimate the number of free adjustable parameters, by counting the degrees of freedom when constructing a matrix having mutually orthogonal normalized columns. In the counting process, the normalization condition is considered independent to the orthogonality condition. As Wax and Kailath deal with a complex covariance matrix having real eigenvalues and complex eigenvectors this yields

$$\kappa_{\mathrm{WK}} = k + 1 + N_{\mathrm{WK}}(\{\vec{v}_i\}_{i=1}^{k}) = k(2p-k)+1. \tag{4.30}$$

However, if we employ the Stiefel-Manifold of complex valued unitary $(p \times k)$ matrices $\mathrm{V}_k(\mathbb{C}^p)$ we obtain

$$\tilde{N}_{\mathrm{WK}}(\{\vec{v}_i\}_{i=1}^{k}) = \dim \mathrm{V}_k\{\mathbb{C}^p\} = 2pk - k^2 \tag{4.31}$$

and thus

$$\tilde{\kappa}_{\mathrm{WK}} = k + 1 + \tilde{N}_{\mathrm{WK}}(\{\vec{v}_i\}_{i=1}^{k}) = k + 1 + 2pk - k^2. \tag{4.32}$$

Comparing equations 4.32 and 4.30 we observe

$$\kappa_{\mathrm{WK}} - \tilde{\kappa}_{\mathrm{WK}} = -k \tag{4.33}$$

and conclude that equation 4.30 systematically underestimates the number of free parameters in $\vec{\theta}^{(k)}$ by k as the normalization and orthogonality conditions can not be regarded independently in the construction process. As stated before, to this day MDL application in hyperspectral image analysis has been reported twice.

Chang and Du published results using Wax' and Kailath's formulation of MDL in [11]. However, in this publication the MDL was not adapted to the real valued signals of hyperspectral images. Therefore, the model code length term systematically overestimates the required model complexity and consequently underestimates the required model order.

Broadwater published a review of various approaches to model order selection, including results of MDL application in [7]. In this paper, the model code length term was reportedly corrected for application to real valued signals and the associated κ term reads:

$$\kappa_{\mathrm{BW}} = 1 + k(\frac{1}{2} + p - k). \quad (4.34)$$

However, it can be shown that κ_{BW} and consequently the code length term is not a strictly increasing series for $k = 1, \ldots, p$. For $k > \lceil p/2 \rceil$ it can be shown that κ_{BW} is monotonically decreasing.

This is even contradictory to the purpose of the code length term as penalty function, as big models that are associated to large values of k are now favored. Thus, MDL is expected to select the largest model available (i. e. $\vec{\theta}^{(p)}$) and indeed Broadwater states that *'MDL never obtained a minimum over the 169 endmembers [...]'.*

These results are summarized in an example for a hyperspectral observation with $p = 200$ spectral bands and $N \approx 21000$ in figure 4.2. The results show that the model codelength term proposed by Broadwater leads to choice of overly complex models, as it suggests decreasing model code lengths for increasing model complexity in the range $100 < k < 200$. We also note that the difference between the complex-valued model codelength term proposed by Wax and Kailath and the alternative for complex valued signals derived in this section (equation 4.32) is almost negligible in the observed scale. However, the codelength term proposed by Wax and Kailath that was adopted in the paper of Chang and Du, is expected to underestimate the required complexity when applied to real-valued signals.

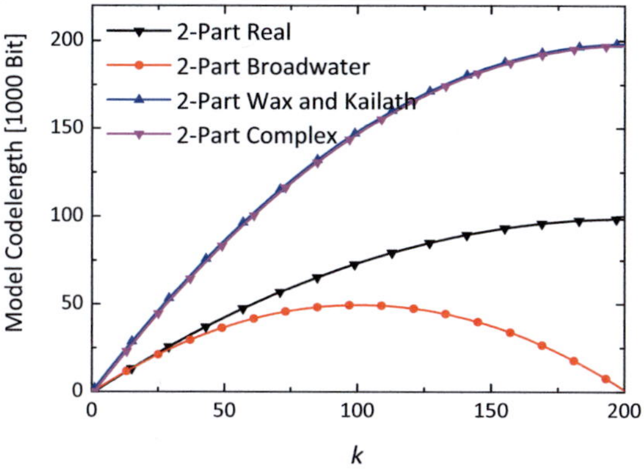

Figure 4.2: Comparison of the model codelength terms found in literature for example of hyperspectral observation with $N = 200$ spectral bands. The model codelength suggested by Wax and Kailath for complex-valued signals is very similar to the result obtained by employing the method of measuring the dimension of the associated Stiefel manifold. Both complex results in general overestimate the model codelength when applied to real-valued signals. The model codelength suggested by Broadwater favors models with high complexity.

We therefore suggest usage of the codelength term given in equation 4.27 for real-valued signals that was derived within this work.

4.4.4 Noise Adjusted MDL

We have up to now restricted ourselves to a data model that accounts for independent, identically distributed (i.i.d) additive white Gaussian noise. I. e. we stipulated that

$$\vec{n} \sim \mathcal{N}(\vec{0}, \sigma^2 \mathrm{I}). \tag{4.35}$$

in the generalized linear mixture model (equation 4.1).

It is a well known fact however that this assumption is often flawed in real-world hyperspectral image data, where correlated noise is commonly observed. Preserving the assumption of a zero mean normal noise distribution, we obtain the slightly more general form

$$\vec{\mathrm{n}} \sim \mathcal{N}(\vec{0}, \mathrm{R}_n). \tag{4.36}$$

We call R_n the *noise covariance matrix.*
Under these conditions, the assumptions concerning the sample covariance matrix eigenvalue distribution made in section 4.1 are obviously violated. Assuming $\vec{\mathrm{n}} \sim \mathcal{N}(\vec{0}, \sigma^2\mathrm{I})$ we were led to believe that the covariance matrix eigenvalue distribution of an observation generated by q linear independent endmembers would satisfy

$$\lambda_1 \geq \cdots \geq \lambda_q \geq \lambda_{q+1} = \cdots = \lambda_p = \sigma^2.$$

This is obviously no longer true for non-i.i.d. noise distributions that follow equation 4.36 and thus renders all following conclusions based upon this observation defective.
The following solution to this problem is based on ideas that lead to the Noise Adjusted Principal Components (NAPC) transform given in [9]. If the noise covariance matrix R_n is known, a *whitening transformation* F can be applied to the observation covariance matrix R, such that

$$\mathrm{F}^\top \mathrm{R}_n \mathrm{F} = \mathrm{I} \text{ and } \mathrm{F}^\top \mathrm{F} = {\Delta_n}^{-1}, \tag{4.37}$$

with Δ_n denoting the diagonal matrix of eigenvalues of R_n. The whitening matrix F is obtained by letting

$$\mathrm{F} = \mathrm{E}{\Delta_n}^{-1/2}, \text{ where } \mathrm{E}^\top \mathrm{R}_n \mathrm{E} = \Delta_n. \tag{4.38}$$

We then obtain a whitened version $\tilde{R}$ of the data covariance matrix R by

$$\tilde{R} = F^\top R F. \tag{4.39}$$

Consequently, we derive a whitened version of the MDL based model order selection method by employing the whitened eigenvalues $\{\tilde{\lambda}_i\}_{i=1}^p$ generated by $\tilde{R}$.

It is left to establish a noise whitening procedure that is applicable if the noise covariance matrix R_n is unknown, in which case it has to be estimated from the available data. As the noise covariance matrix is explicitly known only in rare cases, several estimation methods have been suggested. Chang and Du suggest an approach that follows [53] for application in hyperspectral image analysis.

The method is based upon a decomposition of the inverse of the sample covariance matrix reading

$$R^{-1} = D_{R^{-1}} E_{R^{-1}} D_{R^{-1}}. \tag{4.40}$$

Here, $D_{R^{-1}}$ denotes a diagonal matrix that contains the diagonal elements of R^{-1} and $E_{R^{-1}}$ containing ones as diagonal elements and correlation coefficients otherwise.

The entries of $D_{R^{-1}} = \text{diag}\{\delta_1, \cdots, \delta_p\}$ can be re-written to

$$\delta_i = (\sigma_i^2(1 - r_{p-i}^2))^{-1/2}, \tag{4.41}$$

where r_{p-i} denotes the multiple correlation coefficient of the i-th spectral band to the $p-1$ other bands obtained by multiple regression theory [10]. Following [10] the matrix elements δ_i can therefore be employed to construct a noise covariance estimate as follows:

$$\tilde{R}_n = \text{diag}\{\delta_1^{-2}, \cdots, \delta_p^{-2}\}. \tag{4.42}$$

To obtain a generic whitening process, this result is plugged into the whitening transformation for the case of a known noise covariance matrix given in equation 4.39 to obtain the whitened sample covariance eigenvalue set $\{\tilde{l}_1,...,\tilde{l}_p\}$. Consequently we obtain the Noise Adjusted Minimum Description Length model order estimate:

$$q_{NA-MDL} = \arg\min_k \left[-\frac{N}{2} \left(\sum_{i=1}^{k} \log \tilde{l}_i + \log \left[\sum_{i=k+1}^{p} \frac{\tilde{l}_i}{p-k} \right]^{p-k} \right) + \frac{1}{2}(k + pk - \frac{1}{2}k(k+1)) \log N \right] . \tag{4.43}$$

5 Analysis and Related Work

Standoff detection of explosive substance and precursor residues has been a field of increasingly active research in the last decade and laser based methods have emerged as the most promising approach. In this chapter we will outline the experimental conditions that are induced by the field of application and present an overview over the existing attempts.

5.1 Boundary Conditions of Target Application

In this work we strive to present a fully automatic remote spectroscopy system for detection of explosive residues. Several criteria were considered relevant in real world applications and hence pursued during development.

Eye-Safety

Most standoff detection methods are based on laser spectroscopy. This however, raises the problem of eye-safety, especially since possible applications require open-path operation in public. Nevertheless, fully eye-safe system operation is considered a critical condition.

Stand-Off Capability

Possible application fields include operation in uncontrolled scenarios, e. g. for identification of potentially dangerous substances. Hence, non-contact operation and measurements across several meters should be possible.

Background Interference

The system should be capable of detecting target substances without any prior knowledge and independently of the substrate that it is deposited on.

Ignition Safety

Methods that require high-power or focused lasers can lead to accidental ignition due to the power densities involved.

Detection Performance

Overall detection performance in terms of sensitivity and selectivity should be as high as possible. I. e. the measurement system's cross-sensitivity between dangerous and harmless substances should be low, while threats are detected with high probability.

5.2 Remote Detection Systems

5.2.1 Stand-off Photoacoustic Spectroscopy

Van Neste et al. developed a stand-off spectroscopy device based on the principle of photoacoustic spectroscopy (PAS) that is illustrated in figure 5.1. [61]. A pulsed tunable quantum cascade laser that provides a tuning range of 9.26 µm to 9.8 µm with a spectral resolution of 0.01 nm is employed to illuminate the sample under test. The backscattered radiation is collected by a gold coated mirror and focused between the prongs of a quartz crystal tuning fork. The laser pulse frequency was set to match the mechanical resonance frequency of the tuning fork, giving rise to an acoustic wave at its air-surface interface. The mechanical displacement is measured as piezoelectric voltage and amplified using lock-in technique.
Van Neste et al. reported measurement of various explosive spectra including TNT, RDX and PETN that match the corresponding Fourier transform ab-

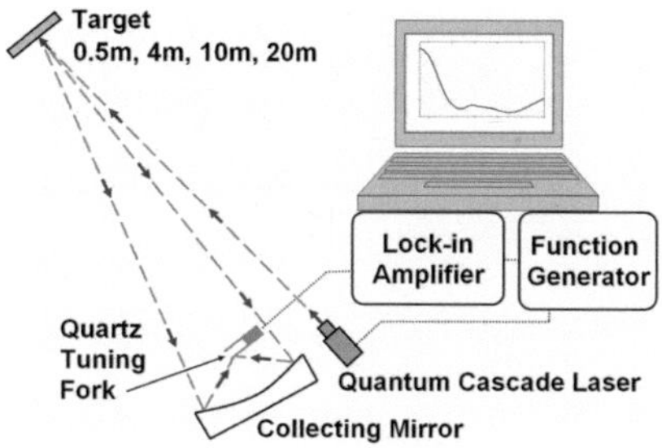

Figure 5.1: Principle of stand-off photoacoustic spectroscopy [60].

sorption spectra well, in the observed measurement range. Measurement results proved predominantly independent of measurement distances between 0.5 and 20 m. Whereas Van Neste et al. provide no further information upon measurement geometry, Kim et al. state in [37] , that *'[...] all these experiments were performed for specularly aligned samples on very smooth surfaces'*. In later publications [37, 40, 60] the quartz crystal tuning fork was replaced by an Mercury Cadmium Telluride (MCT) single element detector. In this configuration the measurement setup can be considered a non-imaging variant to that presented within this work. This comes at the loss of spatial information and renders the sub-pixel target detection methods inapplicable, as the background spectral interference can not be extracted from the measurement data.

In [40] the system is equipped with a UV laser, that is employed to change the target signature between two measurements and hence allow to eliminate background interference for certain target substance molecules that are structurally sensitive to UV light (e. g. TNT). [37] choose a Daylight Solutions MIRcat^{TM} as wavelength selective illumination source, that contains four Quantum Cascade Lasers (QCLs) of adjacent tuning ranges and provides a broad cumulative tuning range from 6.3 µm to 12.5 µm (1587 cm^{-1} to 800 cm^{-1}). In this publication first experimental results in non-specular geometry were presented. As the signal showed less convincing matches to

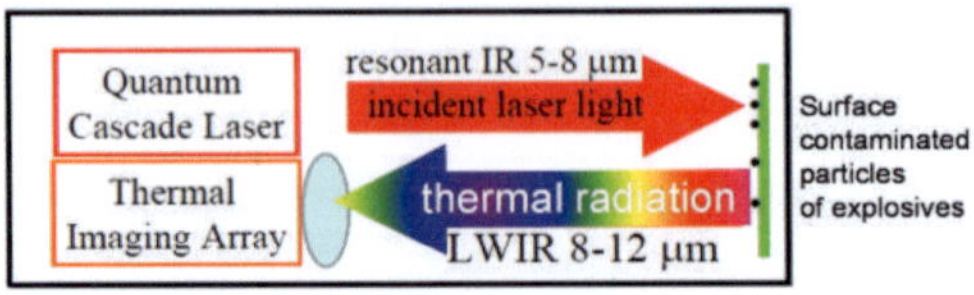

Figure 5.2: Principle of photothermal infrared imaging spectroscopy [35].

the expected library spectrum, the authors closed with the remark, that although *'[...] standoff detection for specularly aligned target explosive samples at large distance was achieved in our previous work, standoff explosive sensing with diffuse reflection spectra seems to be more challenging and needs to be further investigated'* [37].

5.2.2 Photothermal Infrared Imaging Spectroscopy

The measurement principle of photothermal infrared imaging also employs tunable infrared quantum cascade lasers as spectrally selective illumination sources. It was suggested by Furstenberg et al. for remote detection of explosives and has since been subject to several publications [18–20, 34–36]. The concept of photothermal infrared imaging is shown in figure 5.2. The principle is based on measurement of wavelength-dependent resonant thermal heating effects, rather than the reflection or scattering of incident radiation, itself. At wavelengths where the excitation radiation is absorbed, the sample is heated and thus emits thermal infrared radiation in a broad spectral range. An infrared sensor, insensitive to the illumination wavelength is used to detect this heating effect.

In [35] a QCL with tuning range between 5.8 μm and 7.7 μm is used as excitation laser. A liquid nitrogen cooled MCT camera equipped with optical filters blocking the excitation wavelength measures the thermal heating effect in the range of 8 μm to 12 μm. Different illumination-pulse modes are applied to derive a hyperspectral image with optimum signal to background contrast. A long pulse heating phase of 125 ms is followed by a 250 ms pause to

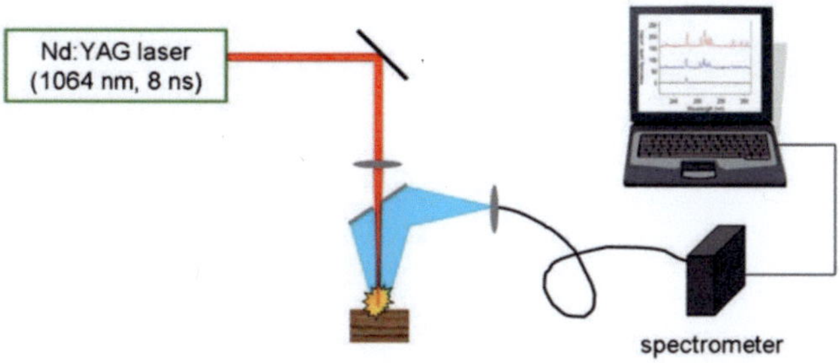

Figure 5.3: Principle of laser induced breakdown spectroscopy [23].

ensure thermal equilibrium is achieved. A series of 16 short pulses < 10 ms completes the sequence, creating a feature vector that shows high discriminative power between target and background. Using this setup, Kendziora et al. report measurement of RDX spectra on several substrates. An ensemble learning algorithm (GentleBoost) is trained on labeled training data and adopted for automatic target detection. Results on a test set suggest an expected probability of detection of 88.3 % for the case of RDX on two different substrates with an expected false alarm rate of 6.0 %.

5.2.3 Laser Induced Breakdown Spectroscopy

Laser induced break down spectroscopy (LIBS) is a variant of atomic emission spectroscopy that is widely used for chemical analyzes. Short pulses of high-power, near-infrared laser light are generated by a Nd:YAG solid state laser and are used to create a plasma in vincinity of the substance to be analyzed. This is achieved by focusing a collimated beam on the sample. During the plasma's cooling process a chemically selective material spectrum can be obtained using a broadband spectrometer (100 nm to 1000 nm). Figure 5.3 illustrates the principle.

Gottfried et al. suggest a trace explosive detection system based on LIBS and report some impressive results for detection of common explosives

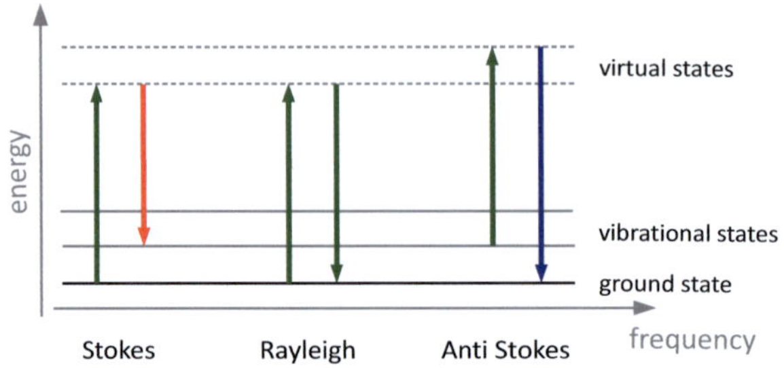

Figure 5.4: High energy incident monochromatic radiation gives rise to Stokes, Rayleigh or anti-Stokes scattering. The energy of a scattered photon can be either identical to the excitation energy (Rayleigh), lower (red-shift / Stokes) or higher (blue-shift / anti-Stokes). The corresponding quantized wavelength shifts are specific to the molecular structure and observed in the Raman spectrum.

like TNT, RDX and PETN using a mobile, short-distance measurement device [23]. However, due to the high power densities required, LIBS is prone to accidental ignition. Gottfried et al. states: *'Ignition (i. e. burning) of an explosive residue has occasionally been observed in our laboratory (primarily with TNT particles), but initiation of an energetic material with a LIBS laser has not been observed for the secondary explosives commonly used by the military'* [23]. Also due to the laser power involved, LIBS is in general not eye safe as the authors state: *'While the LIBS laser will never be completely safe in the direct path of the focused beam, diffuse reflections will be less of a problem at some wavelengths'*. Hence, open path operation requires extensive precautions like securing the beam path.

5.2.4 Stand-off Imaging Raman Spectroscopy

Raman spectroscopy is a well established laser based spectroscopy method, that exploits inelastic light scattering from sample molecules which change their vibrational state during this scattering event [13].
A Raman spectrum is obtained by illuminating the sample with a NIR, visible or UV laser and the scattered radiation is analyzed by a spectrometer. The

laser excitation gives rise to three different scattering effects, that differ by molecular rotational and vibrational energy transitions illustrated in figure 5.4: Stokes scattering, Rayleigh scattering and anti-Stokes scattering. The excitation radiation causes molecular energy transitions to virtual energy levels (dashed lines) that are instable, causing the molecule to immediately relax into one of the indicated low vibrational or ground energy states. In Rayleigh scattering – the by far strongest component of scattered radiation – the molecule falls back into its initial state and hence, the scattered photon has the same wavelength as the excitation radiation. With less probability the molecule falls back into an energy state, that differs from its original one. While in the Stokes Raman process a vibrational mode of the molecule is excited, i. e. can occur at any temperature, the Anti-Stokes process requires the molecule to be already in an excited vibrational state, i. e. requires thermal excitation of the molecule. Taking Stokes-Raman scattering as the most likely process the difference between incident and scattered photon energy equals the energy of the vibrational mode excited, which in turn is characteristic for the molecular species.

Östmark et al. developed an imaging spectroscopy system for remote detection of hazardous substances based on the Raman effect ([49], [45], [46]). The principle is shown in figure 5.5. A frequency-doubled Nd:YAG laser emitting at 532 nm is used as excitation laser, to illuminate the sample. The backscattered radiation is collected by a 200 mm Smith-Cassegrain telescope, mounted in front of an ICCD camera. An optical notch filter at the entrance aperture of the telescope suppresses the reflected light and light caused by Rayleigh scattering. An optical liquid crystal tunable filter placed between collection optics and camera serves as wavelength selective element, enabling imaging spectroscopy. The sensor generates multispectral images, and custom model-fitting based detection algorithms are employed for target detection. Traces of Ammonium Nitrate, TNT and DNT were detected down to 1 g of material at a distance of 10 m.

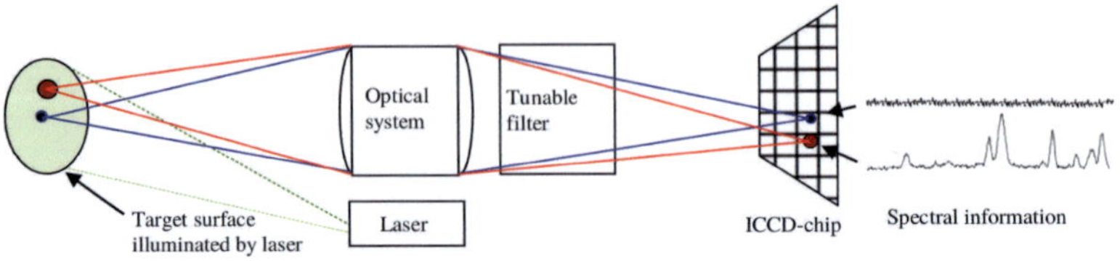

Figure 5.5: The principle of imaging raman spectroscopy proposed by Nordberg et al.. A frequency doubled Nd:YAG laser emitting at 532 nm is used as excitation laser. Rayleigh scatter and specular reflection is surpressed by a notch filter and a tunable liquid crystal filter between collection optics and sensor serves as wavelength selective element [46].

Due to the laser source employed, the system significantly exceeds allowed power density limits for eye-safety which renders open path operation in public areas not possible at this time. In the recent past promising results have been reported towards imaging Raman spectroscopy using eyesafe deep UV excitation lasers [21]. However, to this day the achievable detection accuracy remains insufficient for robust trace detection in the field.

5.2.5 Conclusion

Stand-off trace detection of explosives has been subject to extensive research in the past, but to this day no method has shown to significantly outperform the others. Highly selective and sensitive spectroscopy methods like LIBS and Raman show impressive results in terms of detection limit, but suffer from eye-safety issues. Photothermal infrared imaging spectroscopy is inherently slow, due to required exposition and cool-down time required. Also, thermal coupling between target and background requires complex models or elaborate methods for background interference suppression. Photoacoustic spectroscopy showed to be too insensitive for measurement of the diffuse backscattering signal and in a generic measurement scenario a specular reflection geometry can not always be achieved.

6 Design and Specifications of the Hyperspectral Imaging Sensor

The measurement principle of tunable mid-infrared laser spectroscopy was outlined in more detail in chapter 2. In this section we describe the experimental setup, including the short range hyperspectral image sensor as well as the long range setup, based on the same principle. We introduce the target spectral library that is used for the detection process. Finally, we present measurement results that support application of the linear mixture model and close with a comparison of measurements obtained with Fourier Transform Infrared (FTIR) diffuse reflectance spectroscopy.

6.1 Hyperspectral Image Sensors

The main difference between the sensor variants are the illumination and optics required for the respective operation distances. Both variants of the hyperspectral image sensor employ tunable External Cavity Quantum Cascade Laser (EC-QCL) to illuminate the sample to be analyzed and collect the diffusely backscattered light by a Mercury Cadmium Telluride (MCT) camera. Hence, these core elements and their following specifications are shared between the systems.

6.1.1 EC-QCL Illumination Source

A dual core EC-QCL illumination source is employed for spectrally selective illumination providing a cumulative spectral tuning range of 1000 cm^{-1} to 1400 cm^{-1}. The Quantum Cascade Laser (QCL) chips were designed, grown

and processed at Fraunhofer IAF with the heterocascading design being optimized for broad emission range (section 2.2.1). The laser chips are operated at room temperature and indium-soldered on custom heat-sinks. Temperature stabilization is realized using Peltier cooling elements.
The QCLs are operated in pulsed mode at 1.7 MHz and 100 ns pulse width, which amounts to 17 % duty cycle. The average power at the center wavelengths under these conditions is typically around 50 mW. Under these operation conditions (pulsed mode in external cavity) the emission linewidth is $\approx 1\,\text{cm}^{-1}$. The collimated output beams are aligned by a custom-coated dichroitic mirror with a cutoff wavelength centered between the QCL chips' center wavelengths. Wavelength tuning is achieved by operating the chips in a Littrow external cavity configuration using a blazed diffraction grating with a grating constant of $150\,\text{mm}^{-1}$ for both lasers. In this setup, the emission wavelength is determined by variation of the grating angle. For angular positioning, the gratings are mounted on piezo-driven rotary positioners that offer an angular resolution of 200 µ°. This theoretically yields a spectral wavelength selection resolution of $0.003\,\text{cm}^{-1}$ at $1000\,\text{cm}^{-1}$ and a resolution of $0.006\,\text{cm}^{-1}$ at $1300\,\text{cm}^{-1}$ respectively, which resides well beyond the typical emission bandwidth in pulsed QCLs ($\approx 1\,\text{cm}^{-1}$).

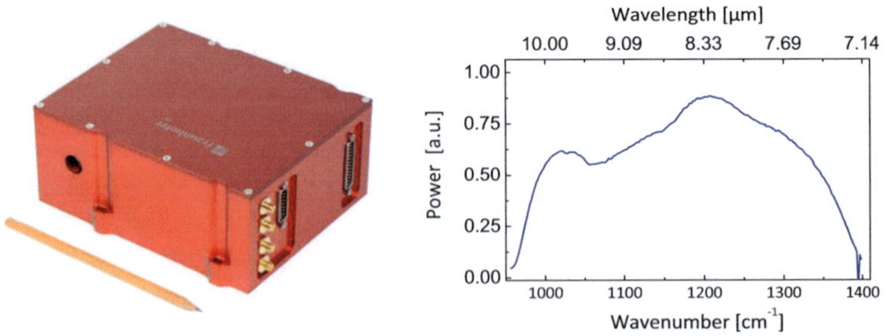

Figure 6.1: Left: dual core EC-QCL module manufactured at Fraunhofer IAF containing two EC-QCLs with adjacent tuning range. Right: cumulative laser tuning curve covering range from $980\,\text{cm}^{-1}$ to $1380\,\text{cm}^{-1}$.

The rotation stage's maximum rotation velocity is 15 °/s, which yields a wavenumber step time of 4.32 ms at the long-wave fringe of the spectrum. The full spectral range can be swept in 0.89 s. The module containing the QCLs is shown in figure 6.1 together with the cumulative laser tuning curve.

6.1.2 Image Sensor

A Stirling cooled MCT camera manufactured by IRCAM is used as infrared image sensor. The detector features 384×288 pixels with a pixel pitch of 24 µm in full frame mode, in which it can be operated at a maximum frame-rate of 160 images per second. To increase measurement speed, the camera is driven in a sub-frame mode of 192×192 pixels, where a frame-rate of 400 images per seconds is achieved. The analog digital conversion resolution is 14 bit. The detector's nominal 20 % cut-on is specified at 7.5 µm (1333 cm^{-1}), long-wave 20 % cut-off is given at 11.5 µm (862 cm^{-1}). The percentage of bad pixels (i. e. pixels that exceed the 2 σ noise variance threshold) was measured to be 0.1 %.

6.1.3 Collection Optics

The short-range hyperspectral image sensor shown in figure 6.2 is equipped with a 100 mm focal length Germanium collection lens with an f-number of $f\,/\,2$. At the illumination laser's spectral long-wave margin of 10 µm, this yields an airy disk of $d_{airy} \approx$ 50 µm and consequently an optical resolution limit of $d_{airy}/2 \approx$ 25 µm. This represents a close matches to the image sensor's pixel pitch of 24 µm.

A modified astronomical Schmidt-Cassegrain telescope with gold coated mirrors serves as collection optics for the long range hyperspectral image sensor (figure 6.3). The telescope's secondary convex mirror was replaced by a plain mirror. The optical characteristics are thus determined by the primary mirror that has a focal length of 1 m and an f-number of $f\,/\,3$.

Figure 6.2: The short range hyperspectral image sensor can be operated between 20 cm and 5 m and provides a spectral tuning range of 1000 cm^{-1} to 1300 cm^{-1}.

Due to the resulting airy disk of $d_{airy} \approx 73$ µm, the optical resolution limit is $d_{airy}/2 \approx 36$ µm, hence the detector slightly over-samples the image signal. Both hyperspectral image sensors are equipped with CMOS cameras, aligned to the infrared collection optics that provide a calibrated visible image of the scene at a large field of view, along with the hyperspectral image. Additional equipment like a distance sensor, a GPS sensor and a pan and tilt sensor can be activated on demand. The corresponding sensor values are stored in a meta-file and have been used e. g. for integration of the sensor into multi-sensor network environments.

6.2 Data Acquisition

Hyperspectral image data acquisition and pre-processing is predominantly controlled by custom acquisition software that is written in the C++ programming language. Due to the time constants involved, hardware trigger sources are employed for various tasks, e. g. to control synchronization of camera frame trigger and the laser pulse modulation.

Figure 6.3: The long range hyperspectral image sensor is operational in the range of 10 m to 25 m and equipped with a CMOS camera aligned to the collection optics that provides a visible image of the scene along with the hyperspectral image.

6.2.1 Illumination Synchronization

The detector's integration time is intrinsically limited to 100 µs. Longer integration times lead to pixel saturation caused by thermal radiation in the scene. At a framerate of 400 frames per second, this limits the system's effective duty cycle to a maximum of 4 %.

It is therefore reasonable to synchronize the laser pulse trigger signal to the camera frame trigger by restricting the laser illumination to the 100 µs camera integration time. This allows to increase the laser's duty cycle to 30 % which – in spite of QCL self heating – can be tolerated by the QCL chips under un-cooled, room temperature operation conditions only for a short time. The camera's read-out phase is employed for heat dissipation.

In addition, as mentioned in chapter 2.4, the hyperspectral images are generated using difference images. I. e. a thermal reference image is recorded by omitting laser illumination in every second image and the difference image serves as signal. This can also be controlled by appropriate choice of the corresponding system trigger signals. We illustrate the resulting time diagram for laser- and camera-synchronization in figure 6.4.

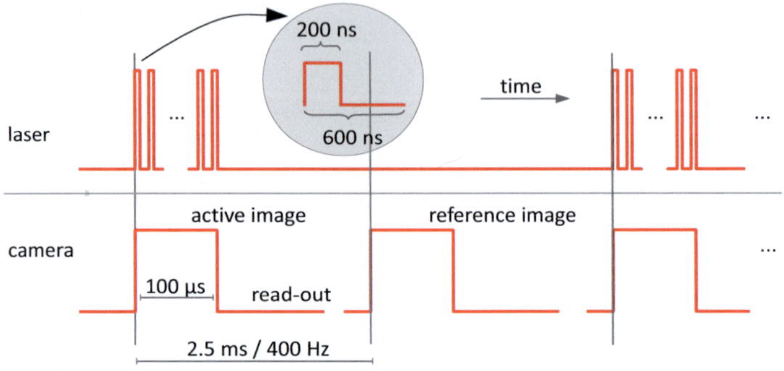

Figure 6.4: Timing diagram for synchronization of laser and camera trigger signals. The high duty cycle laser trigger signal is enveloped by the camera frame trigger that controls the detector's integration time, allowing the chip temperature to stabilize during the detector's read-out phase. Every second image, the illumination is omitted to record a passive image of the thermal scene.

6.2.2 Camera Readout

Synchronization of camera and illumination source is also relevant in implementation of the data acquisition process. We recall that the laser backscattering measurement principle requires difference images and several subsequent difference images are averaged to achieve speckle reduction. To gain reproducible measurements, it is eminent to verify that the emission wavelength is well defined (either constant or scanning over a defined range) during the capturing phase.

In an earlier implementation this was controlled by driving the trigger source that generates the frame trigger in a burst mode. I. e., in the measurement routine, the current emission wavelength was monitored and a software trigger was generated, shortly after the target emission wavelength was reached. The software trigger was fed into the trigger source that generated a sequence of a fixed number of camera and laser pulse trigger signals as described in section 6.2.1. This mode of operation offers the benefit of generating difference images of defined signature – the first image in a burst is always an *active* image and thus thermal background referencing is straight forward. It was stated before that – for a typical spectral resolution of 1 cm^{-1} – the rotation stage that carries the diffraction grating in the external resonator and is thus responsible for emission wavelength selection, requires a step time of up to $\approx$ 4.32 ms, during which no camera readout is performed in the suggested implementation. This mode of camera operation proved highly disadvantageous in terms of image-to-image Signal to Noise Ratio (SNR) variation, as the SNR showed to be directly dependent upon the preceding no-readout time. Readout pauses cause the first image in a burst phase to be subject to significant noise artifacts, and are therefore eminent to prevent.

Hence, the alternative implementation given in listing 4 was developed, which ensures continuous readout. I. e. the software trigger is omitted in favor of a continuous operation of the frame and laser trigger source. Once the target emission wavelength is set, the measurement routine averages the subsequent difference images to determine the image layer to the current illumination wavelength. Whereas it ensures a constant SNR, this implementation comes at the cost of an unknown signal value sign, as at the beginning of each capturing process the relative phase of the synchronized triggers can not be actively set. In the case of strong backscattering signals, the sign of every image layer can be corrected by comparing the signal energy of the averaged image to zero in hindsight. However, if the signal energy is low due to low backscattering radiation or a small illuminated area, this approach may cause severe spectral artifacts.

To elude this issue, the process of continuous frame grabbing was moved into a dedicated image grabbing thread that holds a frame counter. The measurement routine then does not enter the capturing process until the frame counter reaches an even value. This ensures that every averaging process begins either with an active illumination or thermal background image. Hence, the sign is either correct for all image layers, or inverse, allowing the mentioned correction approach to be applicable to the cumulative signal energy of the full image stack.

Listing 4 : Hyperspectral image capturing process

Input : Start and end wavenumbers w_1, w_p, wavenumber step Δw and number of difference images to average per illumination wavelength n_{avg}

Output : Hyperspectral image X

```
GrabbingThread.Start()          // start continuous grabbing thread
X ← ImageStack()                                // initialize result

w_cur ← w_1                             // begin with first wavenumber
while w_cur < w_p do
    Laser.SetEmissionWavenumber(w_cur)          // tune laser to w_cur
    Laser.Join()                          // wait for laser to reach w_1

    while ¬isEven(GrabbingThread.FrameCounter) do
        System.Wait()                     // wait for even frame counter

    Y ← 0                              // temporary image for averaging
    for i ← 1 to n_avg do
        Δ1 ← GrabbingThread.GrabImage()
        Δ2 ← GrabbingThread.GrabImage()
        Y ← Y +(Δ1 − Δ2)/n_avg       // calculate difference images and
          add to Y

    X.Append(Y)         // append mean difference images to result

    w_cur ← w_cur + Δw                            // set next wavenumber

// sum up all values in image cube and check sign
if X.PixelValueSum() < 0 then
    X ← −X

return X
```

6.3 Pre-processing

Several pre-processing steps have to be applied to the acquired hyperspectral images, before they can be employed for further analysis. A spectral normalization procedure renders the data invariant to the laser and coarse sensor characteristics and a homogenization procedure takes care of spectral characteristics on a single pixel level. Finally, defective image pixels are identified and replaced by estimates derived from their neighborhood.

6.3.1 Data Normalization

In section 6.1.1 we showed the spectral emission characteristics of the EC-QCL illumination source (figure 6.1). The emission power is obviously not constant throughout the considered spectral range and consequently, the obtained backscattering radiation is subject to the same variation. Similarly, the sensitivity of the image sensor is subject to spectral variation. Finally, the atmospheric transmitivity is subject to variations caused e. g. by humidity. It is therefore eminent to systematically eliminate these signal contributions, to allow comparability of data obtained with different combinations of laser and camera as well as under other atmospheric conditions.
This is done prior to sample measurement using a normalization procedure. A spectrally flat, sanded aluminum plate with defined roughness serves as normalization sample. The mean spectrum is employed for normalizing all following measurements by element-wise multiplication of every observed spectrum with the inverse of the normalization spectrum.
Let $\mathrm{A} \in \mathbb{R}^{p \times N}$ denote the raw hyperspectral image data acquired using the sanded aluminum plate and $\mathrm{X} \in \mathbb{R}^{p \times M}$ denote the raw backscattering spectra of the sample to be analyzed.
We obtain a representation of the system's transfer function from $\mathrm{A} = [\vec{\mathrm{a}}_1, \ldots, \vec{\mathrm{a}}_N]$ as the vector

$$\vec{\mu} := \frac{1}{N} \sum_{i=1}^{N} \vec{\mathrm{a}}_i \tag{6.1}$$

and – with $\vec{\mu}_{inv}$ denoting the element-wise inverse and $\circ$ denoting the Hadamard product – obtain the normalized hyperspectral image data

$$\tilde{\mathrm{X}} = \{\vec{\mathrm{x}}_i \circ \vec{\mu}_{inv} \,|\, \vec{\mathrm{x}}_i \in \mathrm{X}\}. \tag{6.2}$$

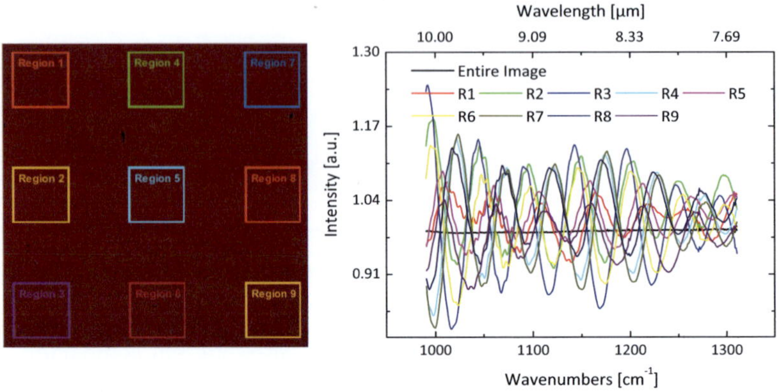

Figure 6.5: Left: definition of averaging regions in hyperspectral image of spectrally flat object. Right: the corresponding spectra after the normalization process still show spatially dependent oscillations attributed to thin-film interference effects in the sensor chip.

6.3.2 Spectral Homogenization

The normalization procedure described in the previous section serves to eliminate the spectral dependent impact of laser illumination, atmospheric transmission and cumulative sensor sensitivity. However, it does not consider non-constant spectral sensitivity differences among the sensor pixels. To evaluate this impact, an experiment was conducted to analyze the homogeneity of the sensor pixels' spectral response to coherent illumination. For this purpose, the laser illumination was projected into a gold coated integrating sphere. The image sensor was used to image the sphere's output port from a distance of 20 cm. Figure 6.5 shows spectra obtained from the hyperspectral image recorded in this configuration after the normalization process. Whereas the mean backscattering spectrum of the entire hyperspectral image is fairly constant, strong oscillations with spatial dependent phase are observed in the averaged spectra of the considered regions. We attribute these to thin-film interference effects at the sensor.

Hence, a homogenization process is required to eliminate this sensor-dependent effect. Let $\mathrm{R} = [\vec{\mathrm{r}}_1, \ldots, \vec{\mathrm{r}}_N]$ denote the hyperspectral data observed using the aforementioned integrating-sphere setup. We obtain a homogenized representation of the hyperspectral image X

$$\tilde{\mathrm{X}} = \mathrm{X} \circ \mathrm{R}_{inv} \tag{6.3}$$

where once more, $\circ$ denotes the Hadamard product and R_{inv} denotes the element-wise inverse of R.

6.3.3 Defective Pixel Correction

A relevant source of noise in images acquired with MCT imaging sensors are defective pixels. Common defects include pixels or pixel clusters that are constantly black (i. e. dark dot defects), constantly bright (i. e. bright dot defects) as well as pixels that switch between the two states rapidly (blinkers). As defective pixel noise is not covered by the data model, it is important to eliminate this noise source in a dedicated pre-processing step. If the locations of the bad pixels are known, it is a valid approach to simply remove the corresponding spectra from the hyperspectral image. However, the missing spectral information introduces an additional noise source if e. g. spatial filters are employed. This issue is eluded by filling the missing pixel values using bilinear interpolation at the bad pixel locations.
As the location of the defective pixels is a constant parameter of the image sensor, it suffices to generate the bad pixel map once for every sensor. This is achieved by a characterization procedure: a series of raw images of a thermally homogeneous object is acquired from which the variance is calculated for each pixel. Pixels that have zero variance are either bright dot defects or dark dot defects and are thus marked as bad pixels. In addition, pixels that show a variance exceeding the 0.95 % quantile are considered blinkers and added to the bad pixel map, too. The effect of this defective pixel removal procedure is illustrated in figure 6.6.

Figure 6.6: Left: logarithm of pixel standard deviation prior to bad pixel replacement. Right: logarithm of pixel standard deviation after bad pixel replacement using bi-linear interpolation. Center: bad pixel map; bright colors indicate presence of a bad pixel.

6.4 The Target Spectral Library

As stated before in chapter 2.3, measurement results acquired by laser backscattering spectroscopy are best comparable to those measured using FTIR spectroscopy. These are well known, to differ significantly from the results that are obtained using Attenuated Total Reflectance (ATR) spectroscopy. Existing commercial and research spectral databases of solids provide the latter and are consequently inapplicable to this task. However, for a set of substances of interest, the target spectral library can be established using the measurement technique on appropriate samples. To obtain pure target substance spectra, signal contributions of interfering substances have to be avoided. This can be achieved by choice of a weakly scattering substrate as carrier for the target substance to be measured – ideally a perfect mirror in a geometry that avoids specular reflection into the detector. Typical carrier substrate choices include e. g. glass or polished metal.
We present the target spectral library containing PETN, RDX, TNT and Ammonium Nitrate obtained from such a procedure in figure 6.7. A weakly scattering piece of painted auto-body served as substrate. The spectra were measured using the short-range hyperspectral image sensor from a distance of 1.4 m. The substances were deposited on the substrate using a silicone stamp with an approximate area of $\approx 1\,\text{cm}^2$. In the measurement configuration, this amounted to ≈ 1000 spectra for averaging per substance.

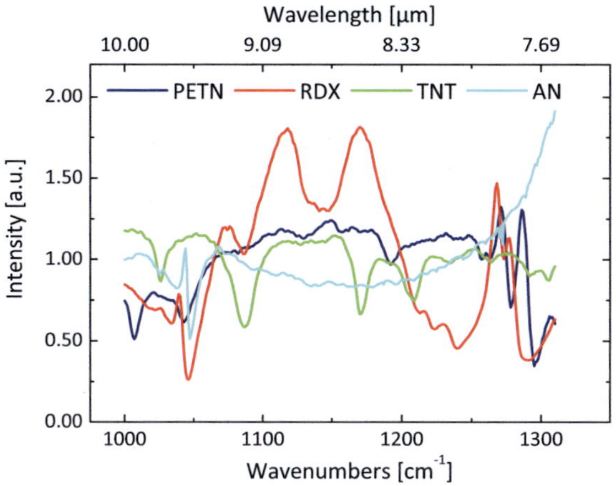

Figure 6.7: Sample traces on a non-scattering surface were measured using laser backscattering spectroscopy to establish a target spectral library.

6.5 Verification of the Linear Mixture Model

If the substance trace to be detected is deposited on a scattering surface, the spectral signature measured by a single pixel may contain contributions of multiple substances. For this reason sub-pixel target detection algorithms usually adopt a data model for spectral unmixing. As stated in section 3.2, a common data model for this purpose in the hyperspectral image analysis community is the Linear Mixture Model (LMM). It is left to be shown that the LMM is applicable for hyperspectral image data acquired by the measurement technique of laser backscattering spectroscopy. In this section we present measurement results that support this assumption.

To verify that the LMM holds, it is sufficient to show that observed spectral mixtures can be generated by linear mixtures of the contributing pure substance spectra. For this purpose, several experiments were performed on homogeneous, but strongly scattering substrates that contain defined contamination regions.

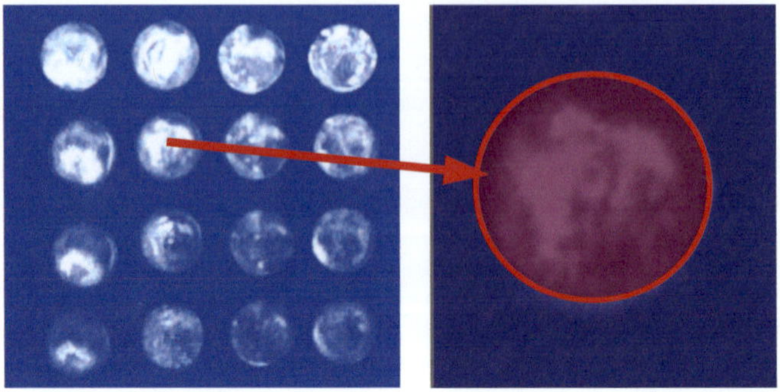

Figure 6.8: Homogeneous Polyamide substrate containing spatially defined contamination areas (contaminant: RDX powder). A pure background spectrum (blue region) and a contaminated spectrum (red region) can be obtained by spatial averaging of the spectra in the indicated regions.

The spectral signature of the contaminant $\vec{t}$ is known in all cases. The defined spatial contamination allows to obtain the mean background signature $\vec{b}$ by spatially averaging of the spectra in the corresponding region. The mean spectrum of the contaminated region $\vec{s}_c$ is obtained from the measurement data, likewise. An example of such a sample is shown in figure 6.8. The homogeneous Polyamide substrate contains spatially defined contamination areas. The contaminant in this case was RDX powder. The mentioned spectra can be obtained by averaging the spectra in the corresponding indicated image regions.
Given LMM holds, we expect that for appropriate choice of abundance coefficients α_t and α_b

$$\vec{s}_c = \alpha_b \cdot \vec{b} + \alpha_t \cdot \vec{t}. \tag{6.4}$$

Hence, the unknown abundance coefficient estimates $\hat{\alpha}_b$ and $\hat{\alpha}_t$ can be achieved by solving the linear equation system

$$[\vec{b}, \vec{t}] \cdot [\hat{\alpha}_b, \hat{\alpha}_t]^\top = \vec{s}_c. \tag{6.5}$$

Typical abundance value constraints on solving this system of two linear equations are the sum-to-one constraint (i. e. $\hat{\alpha}_b + \hat{\alpha}_t = 1$) and the non-negativity constraint (i. e. $\hat{\alpha}_b \geq 0$, $\hat{\alpha}_t \geq 0$). However, the scaling of the measured spectra is in general unknown due to the measurement method. Hence, the sum-to-one constraint is inapplicable and we are left with the non-negativity constraint only.

Having calculated the abundance estimates, we should be able to un-mix an estimate $\hat{\vec{t}}$ of the pure substance spectrum $\vec{t}$ from the observed mixture $\vec{s}_c$ by letting

$$\hat{\vec{t}} = \frac{1}{\hat{\alpha}_t}(\vec{s}_c - \hat{\alpha}_b \vec{b}). \tag{6.6}$$

If the LMM holds, we expect $\hat{\vec{t}}$ to match $\vec{t}$ well. Mathematically, we expect a small residual norm

$$\|[\vec{b}, \vec{t}] \cdot [\hat{\alpha}_b, \hat{\alpha}_t]^\top - \vec{s}_c\|^2 \approx 0. \tag{6.7}$$

The results of the linear unmixing approach for an RDX contaminated Polyamide sample is given in figure 6.9. The hyperspectral image was recorded using the short range hyperspectral image sensor from a distance of $\approx 2\,\mathrm{m}$ with a spectral resolution of $1\,\mathrm{cm}^{-1}$. The pure background and contaminated spectra were obtained by averaging the spectra in the regions indicated in figure 6.8. The spectrum of the contaminated area shows contributions of the contaminant (RDX powder) as well as the background spectrum (Polyamide). Linear unmixing under the non-negativity constraint allows to extract a well matching representation of the RDX library spectrum from the observed spectrum of the contaminated area.

In figure 6.10 we present the linear unmixing residual norm image for two non-matching choices of target spectra in comparison to the correct choice (PETN, TNT and RDX, respectively). For eased comparison to the true substance distribution, the residual norm images are transparently overlaid over a gray-scale visible image of the scene.

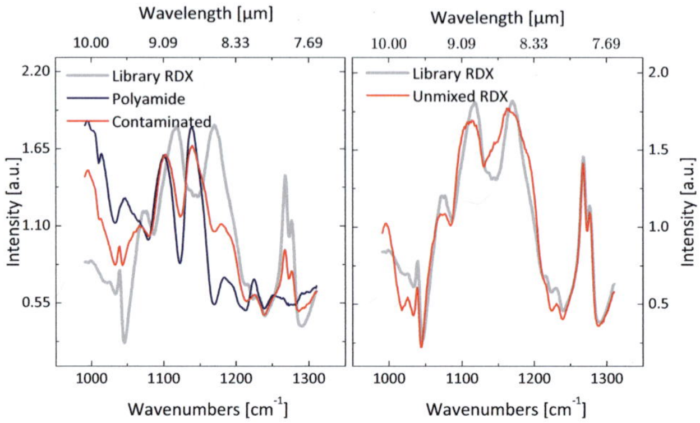

Figure 6.9: Verification of the LMM for the case of RDX contaminated Polyamide.

Keeping in mind that low residual norms indicate good model matches we observe that all models perform equally well in the background region of the hyperspectral image. In the contaminated regions, the PETN and TNT models show poor goodness of fit and as expected the RDX model performs best. We also note that the residual norm image of the RDX model shows no spatial structure. As the substance deposition was not homogeneous within the contaminated areas, this indicates that the goodness of fit is independent to variation of substance density and layer thickness.

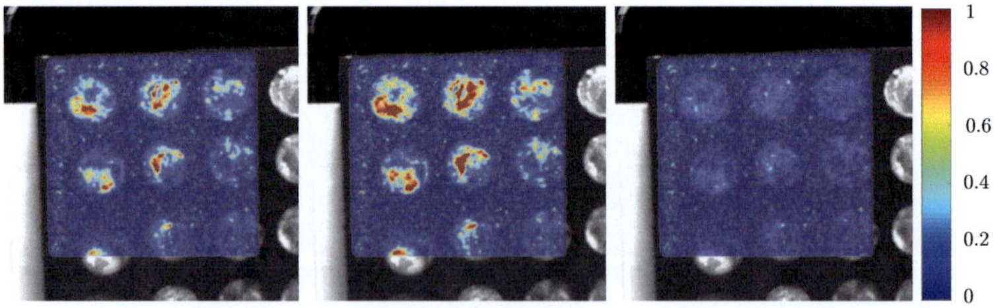

Figure 6.10: Residual norm image of linear unmixing with non-negativity abundance constraint for RDX contaminated Polyamide sample and several possible target spectra (from left to right: PETN, TNT and RDX). A low residual norm indicates a good model match. As expected, the lowest residual norm is obtained adopting a model containing the correct contaminant (RDX).

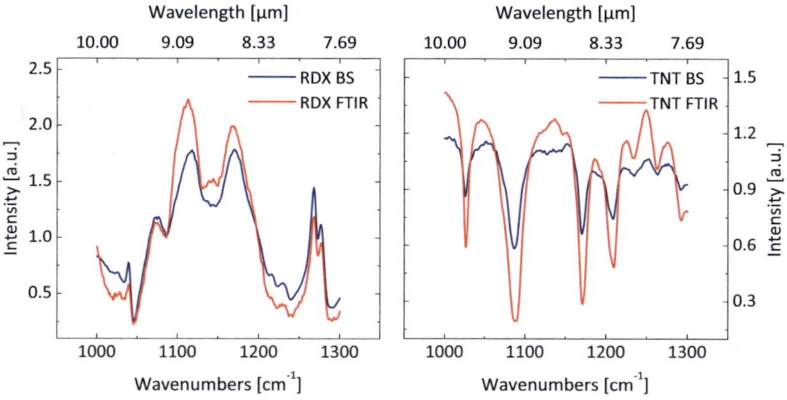

Figure 6.11: Backscattering (BS) spectra in comparison to results obtained by FTIR reflectance spectroscopy. The latter where obtained using a Bruker HYPERION 3000 microscope by averaging over several single grain spectra. The laser backscattering spectra are taken from the substance library. The measurement methods yield very well comparable results.

6.6 Comparison to FTIR Reflectance Spectroscopy

In this section we compare the spectra obtained by laser backscattering spectroscopy to the well established method of FTIR reflectance spectroscopy. Figure 6.11 shows the TNT and RDX library spectra presented in section 6.4 together with the corresponding FTIR reflectance spectra measured with a Bruker HYPERION 3000 FTIR microscope. The FTIR spectra were obtained by averaging of several (≈ 10) single grain spectra. Both measurement methods yield very well comparable results.

These results suggest that the spectral resolution obtained using the EC-QCL is well suited for spectroscopy of solids. We also follow that physical effects like spectral mode-hopping that were mentioned in chapter 2, do not affect the measurement results at the observed scale. These findings support the assumptions and conclusions drawn in section 2.4. We note that in general it is a valid approach to assemble the spectral library for substance detection in hyperspectral images acquired using laser backscattering spectroscopy with FTIR microscope reflectance spectra. This approach bears the benefit that the library spectral range is not limited to the spectral range provided by the laser spectroscopy system and is not prone to noise sources that might affect the measurement results of the latter.

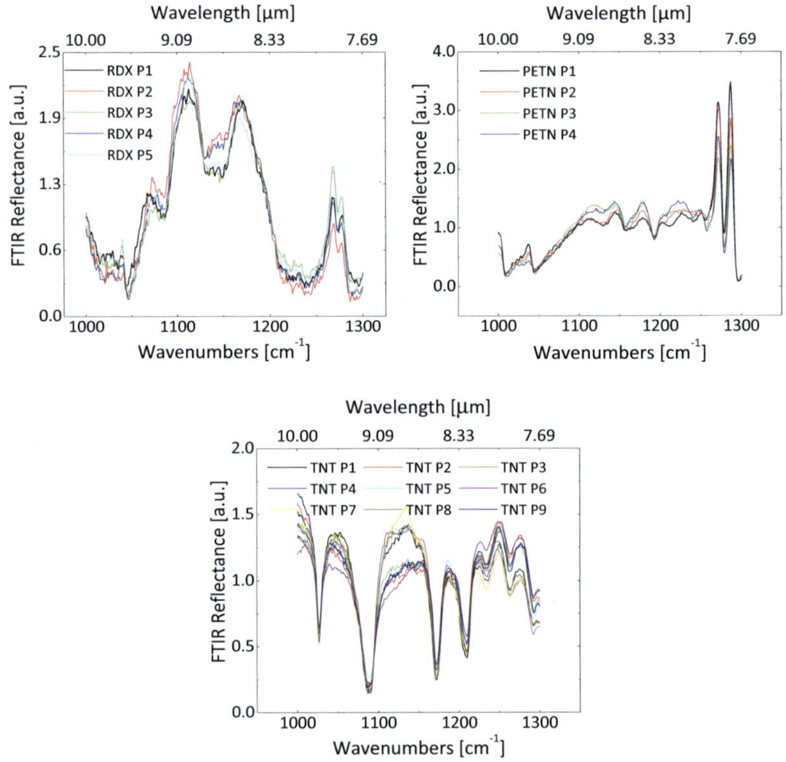

Figure 6.12: Comparison of several FTIR reflectance measurements of single substance grains. The measurements were performed with a Bruker HYPERION 3000 FTIR microscope in reflectance geometry. The characteristic spectral features are well preserved among different grains of the same substance. Slight baseline drifts are observed e. g. in the case of TNT spectra.

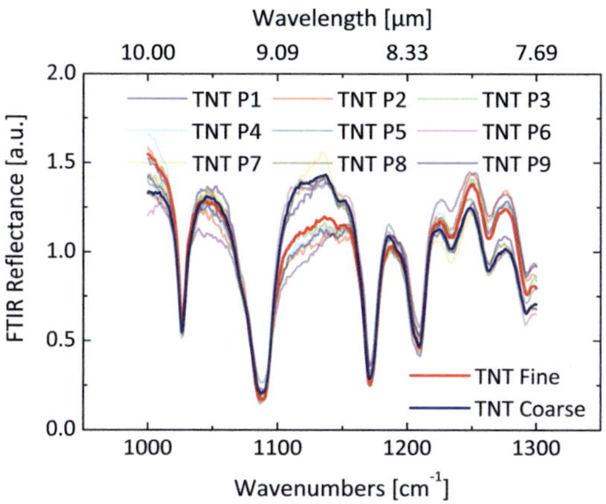

Figure 6.13: Comparison of TNT FTIR reflectance spectra of two different grain sizes (fine and coarse). While the characteristic spectral features of TNT are well observable in both spectra, the grain sizes cause a slight baseline shift.

6.6.1 Intra-Substance Spectral Variability

To obtain an estimate of spectral reproducibility, a series of measurements on single substance grains was conducted to analyze the intra-substance spectral variability. The measurements were performed using a Bruker HYPERION 3000 FTIR microscope.

The results are given in figure 6.12. The material specific characteristic spectral features are obviously shared among several grains of the same substance. Spectral variations within a group of grains are predominantly of low-pass nature and most strongly observed in the case of TNT.

The set of single grain spectra of the latter can be perceived to fall into two disjoint subsets being formed by grains P1 to P4 versus grains P5 to P9. The baseline shift causing the spectral differences is especially strongly pronounced in the range from 1100 cm^{-1} to 1170 cm^{-1} but are also observable at the short wave fringe of the considered spectral range.

Figure 6.13 shows the mean spectra of these two TNT clusters to illustrate the effect in more detail. The observed findings suggest that the spectral differences can be attributed to grain size: grains P1 to P4 were taken from a TNT sample that was subject to a milling technique that yields a finer grained powder during the manufacturing process.

We conclude that the impact of grain size on the observed spectra is not likely to affect the characteristic spectral features of a substance in term of absorption peak positions, but rather introduces a baseline drift of low-pass nature. Consequently – if necessary – such differences can be easily handled by employing a high-pass filter to the spectra in a pre-processing step, or by including spectral variations of the same substance into the library.

7 Analysis and Performance Evaluation

In this chapter we present a detailed comparison of the hyperspectral data analysis methods presented so far under special consideration of specific properties of the proposed active infrared backscattering spectroscopic hyperspectral imaging principle. We begin with an analysis of the Minimum Description Length model order selection method in section 7.2. In section 7.3.1 we illustrate the operation of the Adaptive Background Generation Process algorithm introduced in chapter 4. The impact of the model order selection outcome on target detection performance is evaluated in section 7.3.2. We then compare the expected target detection performance of the considered target detection algorithms in section 7.3.6 and their sensitivity to effects that are specific to the measurement methods (e. g. beam profile and spectral resolution) in section 7.3.5. We close with target detection results obtained on real hyperspectral image data in section 7.4.

7.1 Artificial Hyperspectral Images

Several performance analysis tasks mentioned require labeled data that is not available for real-world data obtained from the presented hyperspectral image sensor. Therefore, artificial hyperspectral images as introduced in section 3.4.1 are employed for this matter.

If not explicitly stated otherwise, the standard hyperspectral image for this task is a 256×256 pixel image that comprises four spatially distinct background regions. The corresponding spectra were measured from real-world samples and represent Polyamide in two different colors, jeans and leather.

The contaminant is deposited in the image center in the form of the Fraunhofer logo with linear decreasing abundance from top to bottom and covers a total number of 1288 pixels.
The general spectral range is set to cover 1000 cm^{-1} to 1300 cm^{-1} at a resolution of 1 cm^{-1}. The illumination beam profile is either top-hat or Gaussian and – using an approximate 4 σ diameter for the latter as threshold – yields 45 561 illuminated pixels. This leads to a contamination coverage of 2.83 %. The wavelength dependence of laser intensity and sensor sensitivity are added using a multiplicative model. The spatial wavelength dependence of the image sensor mentioned in section 6.3.2 is accounted for using the sensor's true spectral responses obtained by measurement of a highly homogeneous and spectrally flat object (a gold coated integrating sphere). The pixel dependent spectral response was added to the simulation data, also using a multiplicative model. If not stated otherwise, Additive White Gaussian Noise (AWGN) is added in a final step, to simulate sensor noise. The standard Signal to Noise Ratio (SNR) is set to 10 dB. Figure 7.1 shows pseudo color representations of the standard artificial hyperspectral image containing the mentioned background regions and a PETN contamination in two beam shape variants. The pseudo-colors were obtained by equidistant binning of adjacent infrared spectral bands into RGB channels.

7.2 Comparison of Model Selection Methods

The goal of model order selection is to estimate the column order of the background endmember matrix, i. e. the parameter q of $\mathrm{S} \in \mathbb{R}^{p \times q}$ in the full Linear Mixture Model (LMM):

$$\vec{\mathrm{x}}_i = \mathrm{S} \cdot \vec{\alpha}_i + \vec{\mathrm{n}}, \mathrm{S} \in \mathbb{R}^{p \times q},\ \vec{\alpha}_i, \vec{\mathrm{n}} \in \mathbb{R}^q. \tag{7.1}$$

In chapter 4 we introduced a parameterless method for this task that is based on the Minimum Description Length (MDL) principle.

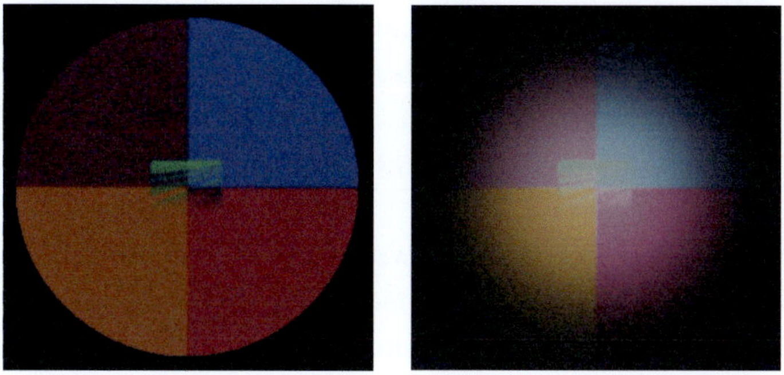

Figure 7.1: Pseudo-color image of artificial hyperspectral image comprising four different background regions and a PETN contamination in the center. The contamination is spread across the center with decreasing abundance from top to bottom. The artificial image was generated with a spectral range from 1000 cm^{-1} to 1300 cm^{-1} and a spectral resolution of 1 cm^{-1}. The top-hat beam shape (left) yields a constant SNR whereas the SNR distribution under a Gaussian beam shape (right) shows a spatial dependence.

In this section we present results of the suggested MDL model order selection method and compare its performance to the Noise Subspace Projection (NSP) method [10] and Second Moment Linear Dimension (SML) methods suggested by Bajorski [4]. As will be subject to a more detailed discussion later in section 7.3.2 the model order is driven mainly by – but needn't necessarily be identical – to the number of spectrally distinct materials in the hyperspectral image. It is enlightening to begin with an analysis of model order selection in artificial hyperspectral image data and consider real-world measurements in a subsequent step.

7.2.1 Impact of Noise Whitening

In chapter 4 we initially derived the MDL based model order selection method based on the LMM in its form given in equation 7.1. In section 4.4.4 we argued that the noise model is restricted to independent identically distributed (i.i.d.) Gaussian white noise. A noise whitening step was suggested leading to the Noise Adjusted Minimum Description Length (NA-

MDL) model selection principle. In this section, we analyze the effect and relevancy of this noise whitening procedure. The aforementioned standard artificial hyperspectral image comprising four different background regions and a PETN contamination serves as hyperspectral observation with known model order for this task.

The noise whitening matrix can be explicitly specified for the case of artificial hyperspectral images. The dewhitening is caused by the normalization stage of the preprocessing step outlined in section 6.3.1 and the whitening matrix can be derived based on this knowledge.

To see this, we recall that during preprocessing the hyperspectral image data $\mathrm{X} = \{\vec{\mathrm{x}}_i\}_{i=1}^N$ is normalized by the element-wise inverse of the mean spectrum of a spectrally flat object $\vec{\mathrm{w}}_{inv}$ using the Hadamard product $\circ$ on each observation spectrum as follows:

$$\tilde{\mathrm{X}} = \{\vec{\mathrm{x}}_i \circ \vec{\mathrm{w}}_{inv} \,|\, \vec{\mathrm{x}}_i \in \mathrm{X}\}. \tag{7.2}$$

We rewrite the normalization procedure using the diagonal matrix W_{inv} that contains the elements of $\vec{\mathrm{w}}_{inv}$ as diagonal elements and obtain:

$$\tilde{\mathrm{X}} = \{\mathrm{W}_{inv} \cdot \vec{\mathrm{x}}_i \,|\, \vec{\mathrm{x}}_i \in \mathrm{X}\}. \tag{7.3}$$

The normalization step affects the linear model of the observation data as follows

$$\tilde{\vec{\mathrm{x}}}_i = \tilde{\mathrm{S}} \cdot \vec{\alpha}_i + \mathrm{W}_{inv} \cdot \vec{\mathrm{n}}_i,\ \vec{\mathrm{n}}_i \sim \mathcal{N}(\vec{0}, \sigma^2 \mathrm{I}). \tag{7.4}$$

Hence, the noise is no longer i.i.d, but rather distributed following

$$\tilde{\vec{\mathrm{n}}}_i \sim \mathcal{N}(\vec{0}, \sigma^2 \vec{\mathrm{W}}_{inv}). \tag{7.5}$$

Using this information we adopt the noise whitening procedure given in section 4.4.4 to obtain the whitened version of the sample covariance matrix required within the NA-MDL based model selection algorithm.

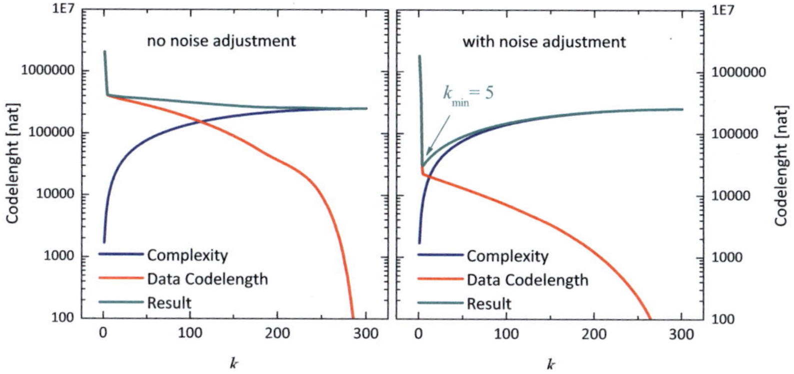

Figure 7.2: Impact of noise whitening step on the outcome of the proposed MDL based model order selection method. The standard artificial image containing five different endmember spectra was employed for this example. Left: the sum of stochastic complexity and data model complexity develops no minimum if noise whitening is omitted. Right: if the noise whitening step is used, the MDL model order selection correctly identifies the expected number of endmembers.

In figure 7.2 we illustrate the effect of noise whitening on the suggested MDL based model order selection method. In the non whitened version the data codelength is significantly overestimated and consequently the result comprised by the sum of model codelength and stochastic complexity does not show a minimum over the potential model orders $1 \leq k \leq p = 300$. If the noise-whitening step with known noise covariance matrix is employed, the NA-MDL yields a significantly reduced data codelength estimate. Consequently, the sum of data codelength and stochastic complexity develops a minimum located at the expected model order index $k = 5$, representing the five different endmember spectra used to generate the artificial image.

7.2.2 Impact of Homogenization

As outlined in section 6.3.2 a crucial preprocessing step is comprised by spectral homogenization, in which spatially dependent spectral pixel characteristics are eliminated.

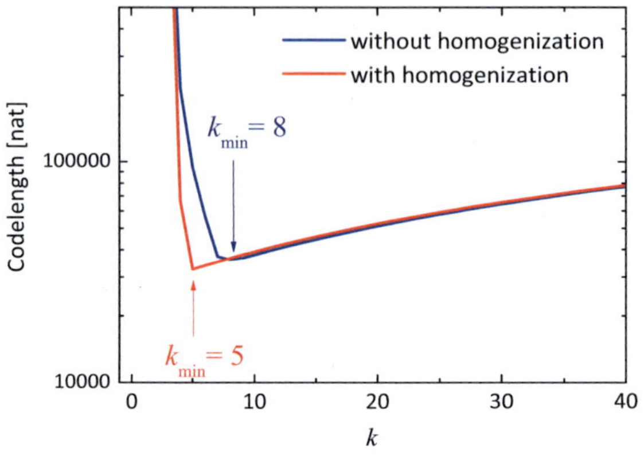

Figure 7.3: Impact of homogenization procedure on MDL based model order selection using artificial data with a true endmember count of $q = 5$. If the homogenization step is omitted the MDL model order selection algorithm overestimates the required model.

The pixel dependent spectral response is attributed to thin-film interference effects within the camera caused by the coherent nature of the illumination and is consequently simulated in artificial images using a multiplicative model. The spatial dependent multiplicative noise source in uncorrected images is not covered by the linear mixture model. It is therefore expected to cause the selected model order estimate to shift towards overly complex models compared to the true number of spectrally distinct materials contained in the scene.

This effect is illustrated in figure 7.3. The corresponding experiment was conducted on a standard artificial image containing $q = 5$ spectrally distinct endmembers. Whereas the MDL model order estimate minimum is blurred and shifted to $k = 8$ in uncorrected data, it shows a considerably more pronounced minimum at the expected value of $k = 5$, if homogenization is applied. This underlines the necessity of the spectral characterization procedure and preprocessing step.

7.2.3 Comparison to Other Methods

The most straightforward way of estimating the model order of a multi-channel observation like a hyperspectral image is to compute the number covariance matrix eigenvalues and threshold their cumulative sum to a fixed fraction of the total eigenvalue sum. This approach has been introduced as Principal Components Analysis (PCA) energy metric in chapter 4.2.1. Chang and Du's NSP [10] and Bajorski's SML [4] comprise more sophisticated methods for evaluating the required model order tailored to application in hyperspectral images. In this section we analyze their accuracy under various circumstances in comparison to the proposed MDL approach using artificial hyperspectral images.

In table 7.1 we present the impact of additive white Gaussian noise on model selection accuracy of the considered methods. We observe that with decreasing SNR the PCA energy metric significantly overestimates the model order. This is caused by the fact that the noise eigenvalues comprise a significant fraction of the total eigenvalue sum. In applications, where a measurement independent fixed SNR can be determined, an appropriate threshold can be established in a calibration step. This is however not a feasible approach in the target application, as the SNR is expected to be subject to measurement specific circumstances (e. g. tilt angle or distance). Hence, we consider the method inapplicable for the target application.

	20 dB	13 dB	10 dB	7 dB	5.2 dB	3 dB	0 dB	−0.8 dB
PCA	4	144	235	263	268	271	272	273
NSP	5	5	**5**	4	4	4	2	1
SML	5	5	4	4	4	4	1	1
MDL	5	5	5	4	4	4	**4**	**3**

Table 7.1: Impact of additive white Gaussian noise on model order selection accuracy.

NSP yields stable results down to SNR levels of 3 dB. Under lower SNR levels the model order is severely underestimated. SML is also reliable for noise levels down to 3 dB from which point on it is essentially dysfunctional. The MDL based model order selection method outperforms the competing methods especially for low SNR. The true model order is underestimated by one even at a SNR of 0 dB.

In table 7.2 we present the considered methods' dependencies on the available spectral resolution of the hyperspectral observation to be analyzed. The experiments were performed on a standard hyperspectral image with a spectral resolution of 1 cm^{-1} in the spectral range from 1100 cm^{-1} to 1300 cm^{-1}, which amounts to $p = 300$ spectral bands. The reduced resolution images were obtained from this image by iterative spectral binning of adjacent bands. For this experiment, the PCA energy metric fraction cutoff threshold was calibrated to yield the correct model order of $q = 5$ under the standard noise conditions (SNR: 10 dB AWGN). With decreasing spectral resolution (i. e. less spectral bands) the PCA energy metric significantly underestimates the true model order. The SML and NSP are in this case well comparable to the PCA energy metric method but are under several other resolution conditions outperformed by the MDL based method. The latter underestimates the true model order only by one, even if only six spectral bands are available for analysis.

	$p = 300$	$p = 150$	$p = 60$	$p = 30$	$p = 12$	$p = 6$
PCA	5	4	4	4	3	2
NSP	5	4	4	4	4	2
SML	4	4	4	4	1	1
MDL	5	**5**	**5**	4	4	**4**

Table 7.2: Model order selection under various resolution conditions. The experiments were performed on the same artificial image in various resolution levels. The original resolution of 1 cm^{-1} yields $p = 300$ spectral bands. The lower resolution images were obtained by spectral binning.

7.2.4 Further Results

To illustrate operation of the MDL model order selection method for various endmember counts, a series of experiments was conducted on artificial hyperspectral images comprising one through ten spectrally distinct endmembers. A geometric distribution of the spectrally distinct materials into four different adjacent regions was employed for the artificial image in figure 7.1. This approach is also used for the case of artificial images containing exactly two background materials. All other configurations were obtained by randomly distributing the contributing spectra across the image. In addition, a spatial low-pass filter is applied to simulate defocused imaging. This causes blurred fringes in the cases of two or four background regions. For all other cases it induces a random mixture of the contributing substances that are unlikely to occur in their pure forms within the resulting images. The MDL model order selection estimations from this experiment are presented in figure 7.4.

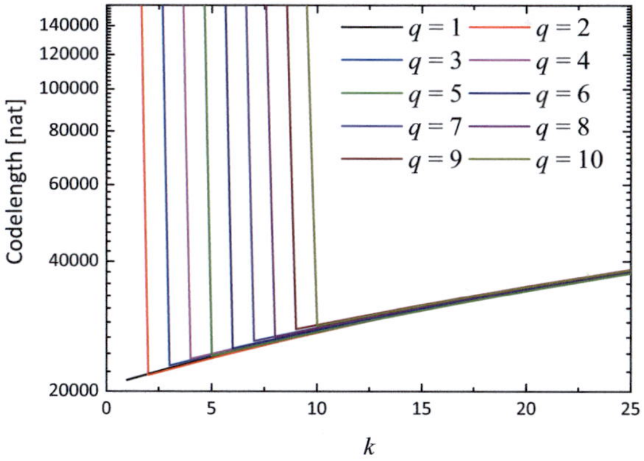

Figure 7.4: Analysis of NA-MDL result for various numbers of spectrally distinct endmember substances. As expected, the minimum shifts by one for every endmember added to the artificial image.

As can be seen, the minimum shifts by one, for every material spectrum that is added to the observation data. No differences can be observed between the two aforementioned fundamentally different background region constellations.

Table 7.3 shows a comparison of model selection results to the other considered approaches. As shown before, the MDL based model order selection method recognizes the underlying model order even in those cases, where the background spectra are randomly arranged and hence, pure background spectra are not contained in the observation data. Among the competing methods, only NSP was able to identify the correct model order with the exception of $q = 9$, under these conditions. The PCA energy metric method was essentially dysfunctional. SML showed to yield systematically underestimated model order estimates for $q = 8$ and higher.

	$q=2$	$q=3$	$q=4$	$q=5$	$q=6$	$q=7$	$q=8$	$q=9$	$q=10$
PCA	1	2	2	3	3	3	3	3	3
NSP	2	3	4	5	6	7	8	8	10
SML	2	3	4	5	6	7	6	6	7
MDL	2	3	4	5	6	7	8	9	10

Table 7.3: Model order selection under various background complexity conditions. The experiments were performed on artificial images containing different numbers of background spectra in various spatial arrangements.

7.3 Target Detection Performance

In this section we present a detailed analysis of the system's detection performance using the considered detection algorithms and background estimation methods under measurement method specific boundary conditions. As stated before, labeled image data is required for estimation of detection performance. This is however not available in the case of real-world hyperspectral images nor is it feasible to acquire.

We therefore conduct the experiments on artificial images and reevaluate the observations on real measurement data in the following section.

7.3.1 Adaptive Background Generation Process

In section 3.5 we presented the Adaptive Background Generation Process (ABGP) algorithm for the task of background endmember extraction as a variant of Ren and Chang's Adaptive Target Generation Process (ATGP). The purpose of background endmember extraction algorithms is to derive a set of linearly independent spectra that span the background subspace of the hyperspectral image. This set of spectra forms the background endmember matrix $\mathrm{S}_B \in \mathbb{R}^{p\times q}$ in the full linear mixture model

$$\vec{\mathrm{x}} = \mathrm{S}_B\vec{\alpha}_B + \mathrm{S}_T\vec{\alpha}_T. \tag{7.6}$$

Background subspace estimation is hence an approach relevant for and shared among all *structured* target detection algorithms based on the LMM. By definition, image spectra are considered background only, if they do not contain the target spectrum. I. e. an individual background endmember set has to be generated for each target spectrum hypothesis in the target spectral library in the detection process.

The ABGP algorithm comprises an iterative approach to background subspace estimation that employs the orthogonal subspace projection operator for operation. The latter is used in the *seeding stage* of the algorithm to generate a mapping of the observation spectra into the subset orthogonal to the concatenation of target spectrum and all background spectra identified so far in each iteration step. The mapped spectrum with the maximum magnitude in this subspace is chosen as the next background endmember and added to the endmember matrix. This process is repeated until the set of q spectra determined by the model selection process is completed.

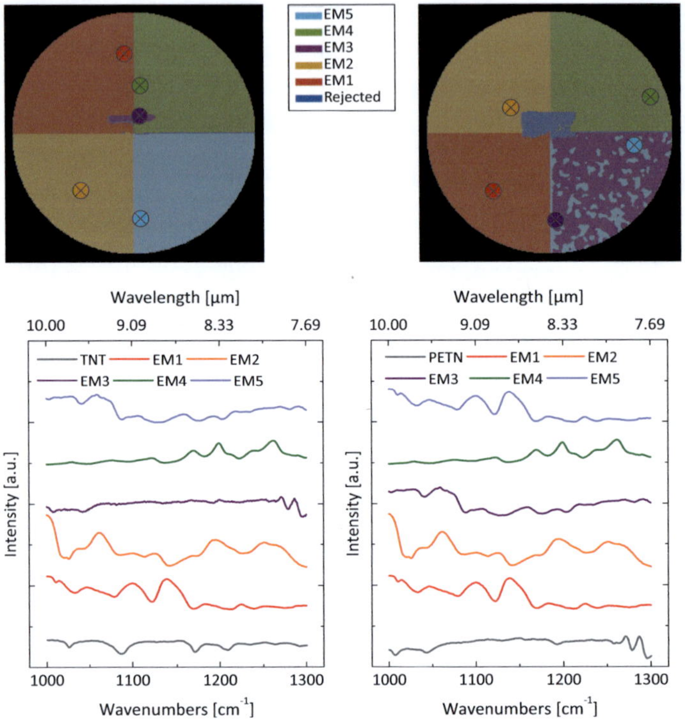

Figure 7.5: The ABGP algorithm operating on artificial hyperspectral image with PETN as target spectrum and four different background regions. Top left: ABGP result for TNT target hypothesis. The endmember locations are indicated by colored circled crosses and the spatial regions containing spectra that form the corresponding cluster are colored accordingly. Top right: same for PETN hypothesis. Bottom: the corresponding endmember spectra formed by the cluster mean vectors.

In a subsequent *stabilization stage* the observation spectra are clustered based on their correlation to the background endmembers and the cluster means are utilized as endmember representatives.

We present the ABGP operation for the example of an artificial hyperspectral image in figure 7.5. The location of the raw endmember spectra chosen in the first part of the process are indicated with a colored, circled cross. The spatial regions containing spectra that form the corresponding cluster are colored accordingly. We present the ABGP outcome for the model order estimation $q = 5$ both for the correct hypothesis of PETN as target spectrum as well as for the competing hypothesis of TNT as target spectrum. In the latter case the ABGP chooses an endmember spectrum for each of the four background regions and an additional endmember in the area with high PETN contamination, as this is considered background under the TNT target hypothesis. Under the PETN target hypothesis, the fifth endmember is obsolete, in the sense that the background subspace order is exceeded. Due to the target inclusion guard, no endmember is chosen from the target contaminated area. Therefore, a redundant background endmember that has no effect on the background subspace is chosen in the area that was already covered by endmember three.

7.3.2 Impact of Model Order Mismatch

In section 7.2 we showed that model selection estimations can potentially be prone to inaccuracies under various noise or low resolution conditions. It is thus important to analyze the impact of model order estimation missmatches upon the target detection performance to be expected.
In figure 7.6 we present the Receiver Operating Characteristics (ROCs) for target detection in a standard artificial hyperspectral image containing four spatially distinct regions and a PETN contamination. The detection results were obtained using the Adaptive Matched Subspace Detector (AMSD) in combination with the ABGP background endmember extraction algorithm for a series of different values of model order estimates. The ROCs show the expected true positive rate in terms of the true false alarm rate for different detection threshold values. We note that false alarms include both false alarms for the target substance as well as false alarms of any other target spectral hypothesis. We observe that if an overly simple model order is selected (i. e. $q < 5$), the detection performance is significantly decreased

in comparison to the detection performance achieved employed by the intuitively correct model order ($q = 5$).
We attribute the quantized, step-function like behavior to the four similarly sized background regions: depending on the threshold value, an underestimated order model falsely classifies one or several background regions that are not represented in the background matrix as contaminated regions. Hence, each region contributes fully or not at all to the false alarm count.
If the model order is overestimated (i. e. $q > 5$) no adverse effect on target detection accuracy is observed. This is true for model orders overestimated by as much as a factor of four. If the model order is overestimated by an order of magnitude, a slight reduction of target detection accuracy is seen. We attribute this minor decrease, to numerical instability of the inverse background endmember matrix that is employed in the AMSD algorithm.

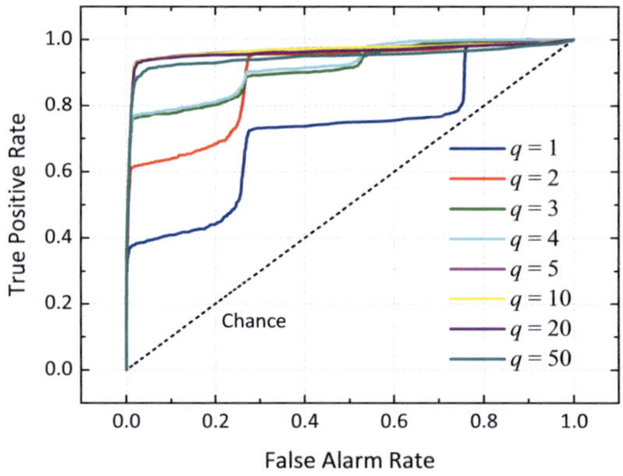

Figure 7.6: The model order estimate determines the number endmembers generated by the background endmember extraction step and affects the detection performance of structured target detection algorithms. If the model order is underestimated, the detection performance is significantly reduced, whereas overestimating the model order has only marginal impact.

From these experiments we conclude that a robust implementation of the ABGP algorithm is well capable of coping with overestimated model orders, whereas significant detection performance reductions are to be expected if the model order is underestimated. Hence, the chosen model order must not necessarily match the true physical number of spectrally distinct materials. However, overly simple models are to be avoided. Nevertheless, as the processing time of the ABGP is linear in sample size but quadratic in model order, it is in general favorable to select an estimate close to the lower margin instead of an arbitrarily high number as value of q.

7.3.3 Spectral Resolution

The target spectral resolution of a hyperspectral image determines the required measurement time as several images at limited frame-rate have to be acquired for every illumination wavelength to generate a spectral band. In addition high spectral resolution images require higher computation power in the analysis step. For these reasons, it is a desirable goal to reduce the measurement resolution without affecting the target detection performance. We present the target detection performance for a series of hyperspectral images of varying spectral resolution in figure 7.7. The experiments were performed on a standard artificial hyperspectral image having a spectral resolution of 1 cm^{-1}. The latter represents the physical spectral resolution limit using an EC-QCL under the specified measurement conditions (see section 6.1.1 for further details). The reduced resolution images were obtained from this image by selection of equidistant spectral channels and dropping the intermediate bands.

The detection performance is compared for five different resolution levels. The highest spectral resolution of 1 cm^{-1}, yields a total number of $p = 300$ spectral bands in the range from 1000 cm^{-1} to 1300 cm^{-1}. Lower resolutions of 2 cm^{-1}, 3 cm^{-1}, and 5 cm^{-1} give rise to $p = 150$, $p = 100$ and $p = 30$ bands. From the results shown in figure 7.7 we conclude that any reduction of spectral resolution affects the resulting detection performance.

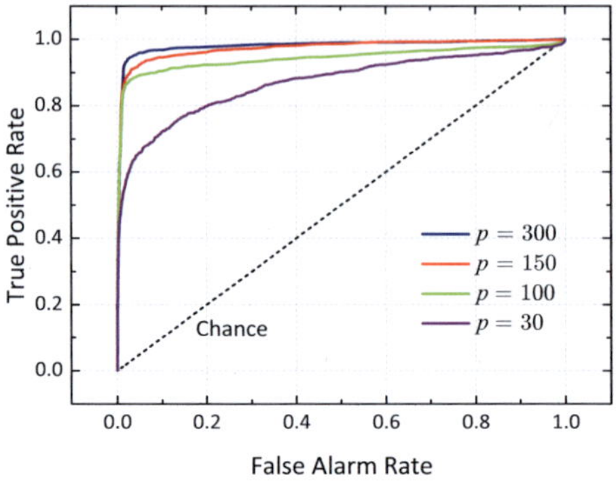

Figure 7.7: Impact of spectral measurement resolution on target detection performance in an artificial hyperspectral image containing a PETN trace on four different background substrates. The maximum spectral resolution of 1 cm^{-1} generates $p = 300$ spectral bands from 1000 cm^{-1} to 1300 cm^{-1}.

The detection performance reduction is marginal between the 1 cm^{-1} and 2 cm^{-1} resolution cases and thus the benefit of improved measurement and analysis computation time might be acceptable.
The discrimination performance drawbacks observed from lower spectral resolution suggest however that further resolution reduction is not advisable. We note, that these results are closely related to the feature characteristics and more specifically to the feature width of the spectra in the considered library. In general, the spectral resolution has to be sufficient to resolve the characteristic features of the target spectra.

7.3.4 Impact of Homogenization Step

An important part in data preprocessing is the homogenization step that was introduced in section 6.3.2. The coherent nature of the active illumination showed to cause thin-film interferences in the detector and induce spatially dependent oscillations on the resulting spectra. This systematic multiplicative noise source is in general not covered by the LMM and is therefore expected to affect the resulting target detection performance.

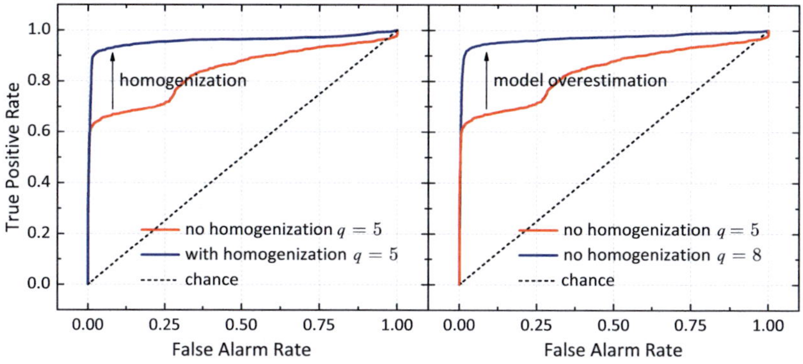

Figure 7.8: Impact of spatially dependent multiplicative oscillations on the measurement results caused by thin-film interference effects in the camera sensor on target detection performance: the significantly decreased detection performance can be either compensated by the homogenization step in data preprocessing (left) or by choosing the systematically increased model order outcome generated by the MDL estimation method (right).

We present the effect of the homogenization step on target detection performance in figure 7.8. The performance results were obtained on a standard artificial hyperspectral image containing a PETN trace on four different background materials with, and without the homogenization step. The AMSD target detection algorithm was used in combination with the ABGP background endmember extraction method. If the true number of different materials (i. e. $q = 5$) is employed as model order for the ABGP algorithm the detection performance is strongly affected if the homogenization step is omitted. In section 7.2 we analyzed the impact of homogenization on the MDL model selection estimate and observed that omitting homogenization increases the model order. For the specific artificial hyperspectral image, the MDL generated an estimated model order of $q = 8$ on the raw observation data. It is therefore interesting to analyze detection performance, if this intuitively overestimated model order is applied in the background extraction step. In figure 7.8 we compare the results obtained on the non-homogenized hyperspectral image that was obtained by the model order

suggested by MDL (i. e. $q = 8$) to the intuitive choice of physically distinct materials in the image (i. e. $q = 5$). It is well observable that using the higher model order generates a target detection performance now again comparable to the results obtained after application of the homogenization step. From these observation we conclude that the thin-film interference induced noise source can be dealt with by elimination of the systematic effect using homogenization. We also follow that an alternative way of handling this effect is to accept the increased model order required by the raw data. We hold the first approach for favorable, as it generates true material spectra and requires lower order models which is considered beneficial in most applications. In addition, the spectroscopy method might be applied for a comprehensive spectral analysis in a scene, in which case omitting the homogenization procedure would lead to significant artifacts.
Finally, we observe that under unknown noise conditions, the intuitive physical choice of model order that is based on the true number of spectrally distinct materials in an observation does not necessarily represent the model order that yields optimum target detection performance. In the case of non-homogenized data, the MDL estimate might initially seem flawed, yet the obtained detection performance of the latter is significantly outperformed by that gained, following the MDL suggestion.

7.3.5 Impact of Beam Profile

The hyperspectral image sensor presented in this work is equipped with speckle reduction optics that provides a top-hat beam profile. In section 6.1 we pointed out that this kind of beam profile is favorable compared to a Gaussian beam profile that is obtained, if no beam shaping is done. In this section we present simulation results that serve to support this statement. Figure 7.9 shows a performance comparison under top-hat and Gaussian beam profile illumination conditions between the Matched Filter (MF) and the AMSD target detection algorithm with ABGP background extraction.

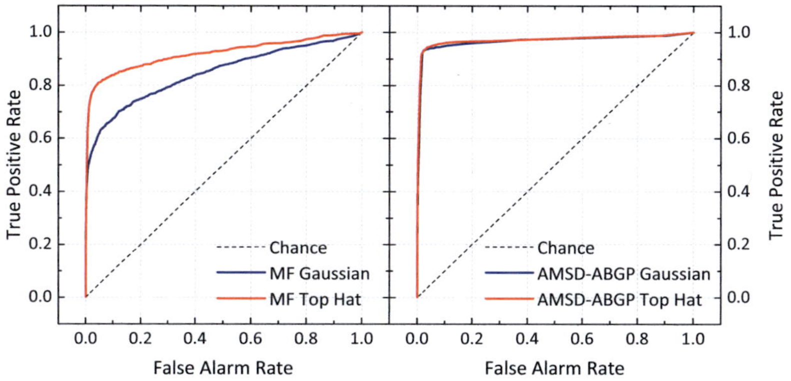

Figure 7.9: Analysis of detection performance on illumination beam profile. Right: the detection performance of the CFAR AMSD detection algorithm is independent to selection of beam profile, whereas the outcome of the non-CFAR MF detector (left) benefits from the constant SNR distribution yielded by the top-hat profile.

Using standard artificial hyperspectral images containing a PETN contaminated area on four background regions for the measurements we observe that the impact of beam profile on detection performance generated by the AMSD detection method is marginal. In contrast, the top-hat beam profile yield significantly better performance results compared to the Gaussian beam profile, if the MF is used for target detection.

These findings are attributed to the Constant False Alarm Rate (CFAR) property of the AMSD algorithm. Under CFAR conditions, the target detection performance is independent to the signal to noise ratio. Consequently, the false alarm rates obtained by the AMSD algorithm using a Gaussian beam profile are spatially independent and thus match the results obtained by choice of a top-hat beam profile. The MF however is not CFAR and thus suffers detection accuracy performance reduction induced by the spatially non-constant SNR distribution.

7.3.6 Comparison of Detection Algorithms

In section 3.3 we presented an overview over target detection algorithms that are widely used in the hyperspectral image analysis community. Based on a hierarchy of two criteria, the detection methods were classified into three families. The family of full pixel target detectors, do not fulfill the sub pixel detection capability criterion and do not rely on a specific data model. This family contains the Normalized Cross Correlation (Normalized Cross Correlation (NCC)), the Matched Filter (MF) and the Constrained Energy Minimization (Constrained Energy Minimization (CEM)) algorithms. Among the sub pixel detectors we differentiate between the structured target detectors (Orthogonal Subspace Projection (OSP) and AMSD) that require a background endmember extraction algorithm and the unstructured methods (e. g. the Adaptive Coherence / Cosine Estimator (ACE)) that model the background as noise clutter.

In this section we compare the target detection performance obtained by the considered methods for the example of the standard artificial hyperspectral image containing a PETN trace spread on four background materials. The AMSD algorithm occurs in two flavors that result from the two considered background endmember extraction methods. The AMSD-PCA generates the background spectra using the sample covariance eigenvectors to the first q most significant eigenvalues (see section 3.5.1). The AMSD implementation that utilizes the proposed ABGP background endmember extraction approach is denoted AMSD-ABGP.

The results are given in figure 7.10. As expected, the sub pixel target detection algorithms significantly outperform all members of the full pixel detectors. This can be contributed to the fact that only few pure target pixels are contained in the image. Among the full pixel detectors the CEM and MF approach yield well comparable detection performance results, while the NCC proves essentially dysfunctional due to exceptionally high false alarm rates, even at high threshold values.

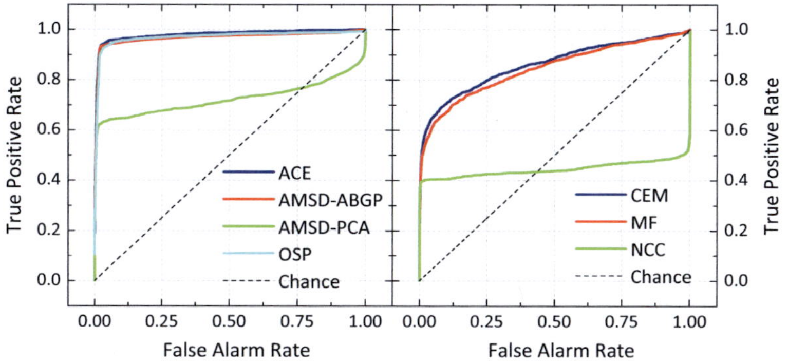

Figure 7.10: Performance analysis of various target detection algorithms. The model-based sub pixel detection algorithms (left) significantly outperform the full pixel detection algorithms (right). No significant difference is observed between structured target detectors (AMSD, OSP) and the unstructured detector (ACE). The ABGP implementation outperforms the more straightforward PCA implementation of the AMSD algorithm.

Among the AMSD implementations, the variant that uses the ABGP endmember extraction method significantly outperforms the more straightforward AMSD-PCA implementation. No significant difference is observed between the considered structured and unstructured target detection algorithms, as OSP, ACE and AMSD-ABGP detection result performance results are well comparable.

Given the findings from these experiments we will employ the AMSD detector with ABGP background extraction (AMSD-ABGP) for most of the following. Not only showed the expected detection performance to be among the most promising of the considered methods. Moreover, its physically meaningful background endmember extraction method and utilization of the same in the hypothesis testing step renders it a transparent and intuitive detection algorithm for the task at hand.

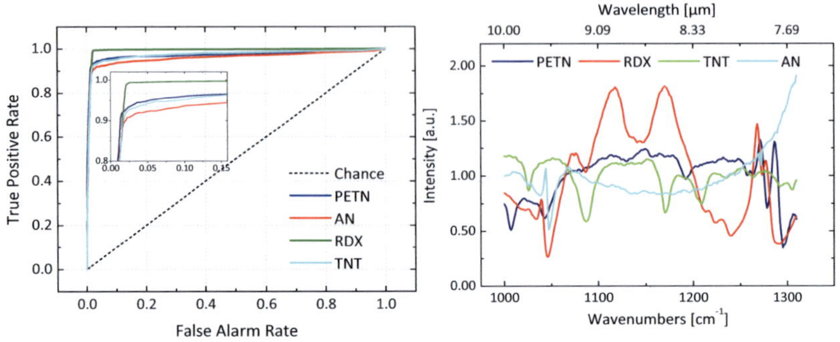

Figure 7.11: Detection performance (left) for the considered spectral library substances (right). Target detection performance is well comparable among the four material spectra, although they show considerable differences in feature richness.

7.3.7 Impact of Substance Spectra

In section 6.4 we presented the library containing the spectra of four hazardous substances that are of interest in many security related applications. While the substance spectra are well distinguishable from among each other, they show considerable differences in spectral feature richness. Whereas PETN, TNT and RDX possess several characteristic spectral features, AN comprises only one characteristic feature at 1042 cm^{-1}. Intuitively, one is led to assume that this renders the latter to be more difficult to detect in comparison to the competing substances. In this section, we seek to investigate, if this notion is supported by experimental data.

A series of experiments was conducted on a set of four artificial hyperspectral images, each containing a trace contamination with one of the substances from the spectral library. The AMSD algorithm with the ABGP background endmember extraction method using a model order of $q = 5$ was employed as target detection process. We present the respective ROCs generated by these configurations in figure 7.11. We observe that at a false alarm rate of 0.5 % a detection performance of greater than 95 % is obtained for every target substance. Nevertheless, slight performance differences can be seen at the lower false alarm fringe. While best results are achieved upon the image containing an RDX contamination, PETN and TNT yield slightly weaker results. As expected, the weakest detection performance among the considered substances is achieved for detection of AN.

7.4 Detection Results on Real World Samples

All results presented in this chapter up to this point were obtained upon artificial hyperspectral image data. Baring these results in mind, we now turn our attention to analysis of real-world measurement data. Various samples containing traces of the considered library substances were available for this analysis. The samples consist of a homogeneous substrate material (e. g. jeans cloth or Polyamide) containing several contaminated areas.
The contaminants were deposited on the substrate using a 1.5 cm diameter round silicone stamp that was dived into the pure substance powder and then placed on the substrate in a regular pattern. Whereas this deposition method does not yield defined quantities of target substance nor a strictly defined spatial distribution, experiments showed, that it serves well to simulate fingerprint traces, expected to be observed in real-world situations. The sample substrates have a side length of 10 cm by 10 cm and typically contain 16 contaminated areas. An example is given in figure 7.12, showing a visible image of the scene transparently overlaid by mean signal energy image and a false color image. The former was obtained by integrating over the spectral dimension of the image, the latter by equidistant spectral binning of adjacent IR channels into RGB channels. The measurements presented in this section were obtained using the short range hyperspectral image sensor presented in chapter 6 from a distance of ≈ 1.4 m. Using a 50 mm projection optics and the rectangular waveguide for beam shaping, the illuminated area at this distance amounts to about $6.4 \times 6.4\ \text{cm}^2$.

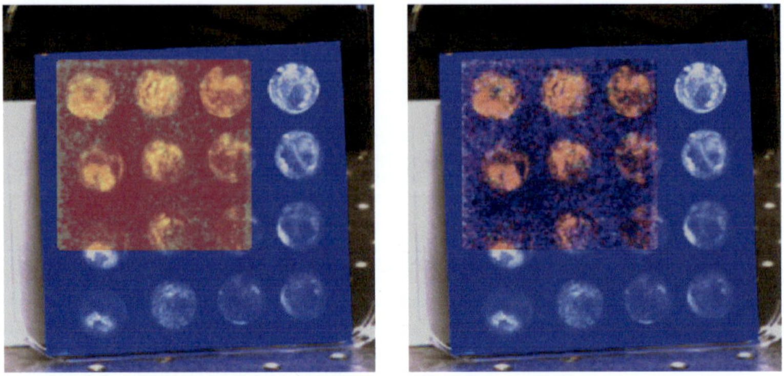

Figure 7.12: Typical sample containing RDX traces on a Polyamide substrate. The visual image is overlaid with the mean backscattering intensity (left) and a pseudo-color representation of the hyperspectral image (right). The pseudo-color representation was obtained by equidistant binning of the IR channels onto RGB channels and indicates spectral differences between target and substrate.

Given the results in section 7.3.3 a spectral resolution of 1 cm^{-1} was chosen in the spectral range from 990 cm^{-1} to 1310 cm^{-1}. Per each illumination wavelength, a total number of 10 difference images was averaged to generate the corresponding spectral band image.
The remainder of this section is organized as follows: we begin with an analysis of the considered model order selection methods on real-world measurement data and follow with an investigation of the ABGP background endmember extraction process. We close with an overview over results obtained by the target detection step for several sample measurements.

7.4.1 Model Order Selection

We presented a comparison of the considered model order selection methods on artificial hyperspectral image data under various noise and resolution conditions in section 7.2. In the following, we will examine, if these results are transferable to real-world measurements as introduced in the

Sample	PCA	NSP	SML	MDL
RDX on jeans	202	53	**2**	71
PETN on jeans	237	44	1	61
TNT on jeans	124	52	1	74
RDX on PA	181	45	**2**	69
PETN on PA	173	50	**2**	81
TNT on PA	142	47	1	75

Table 7.4: Model order estimates generated by the considered estimation methods on various hyperspectral images obtained by the short range hyperspectral image sensor.

beginning of this chapter. We note that all considered samples contain exactly two different substances: the substrate material and the contaminant. In table 7.4 we present the outcome of the PCA, NSP, SML and noise whitened MDL model order selection methods upon various samples. Among the considered methods the results generated by the SML method match most closely the intuitive expected model order based on the physical number of spectrally distinct materials in the image. All other considered methods significantly exceed this number up to two orders of magnitude in the case of the PCA based method. The NSP consistently produces estimates of around 50, whereas the MDL based method generates estimates of 61 to 81. The MDL estimates lie significantly beyond the estimates expected from the simulation given in section 7.2. In a robust implementation of the ABGP, this model overestimation has no impact on the detection performance as shown in section 7.3.2 for the case of artificial data and again later in section 7.4.3 for the case of real measurement data. However, as stated before, the computation time of the ABGP depends predominantly on the target model order, hence lower model order estimates would be beneficial. It is therefore of interest to identify the effect that causes the observed behavior of the MDL and NSP methods on real measurement data.

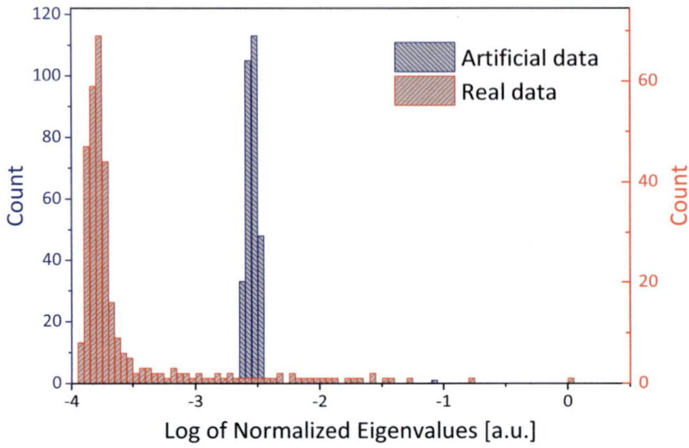

Figure 7.13: Comparison of covariance matrix eigenvalue distributions of artificial and real measurement data containing RDX on jeans substrate. We show the histogram of logarithmic normalized eigenvalues of the respective sample covariance matrices. Whereas eigenvalues generated by the sample covariance matrix of artificial hyperspectral image data are obviously well localized in comparison to the those generated by real data. This effect is attributed to correlated noise observed in the real measurement data.

An indication can be derived from a comparison of the covariance matrix eigenvalue distributions obtained from real and artificial hyperspectral image data as given in figure 7.13. We present the histogram of eigenvalues generated by the sample covariance matrix of an artificial and real hyperspectral image, both containing an RDX contamination on jeans substrate. The eigenvalues were obtained from the noise whitened covariance matrices (see section 4.4.4) and scaled to a maximum of one. Due to the large value range, the histograms were generated on the logarithms of the corresponding data sets.

From the distributions observed in figure 7.13 we conclude that the general noise level was significantly higher in the artificial hyperspectral image. However, in comparison to the eigenvalues obtained from the real measurement, the artificial noise related eigenvalues are more concentrated around

the mean noise variance. The broader distribution of the real measurement covariance matrix eigenvalue indicate presence of correlated noise that was obviously not fully removed by the noise whitening step.
Correlated noise is commonly observed in many hyperspectral imaging applications [53]. The SML method that performed best in terms of being closest to the true physical number of spectrally distinct materials in the scene, is based on an empiric observation on the distribution of covariance matrix eigenvalues in hyperspectral images. It therefore intrinsically accounts for presence of correlated noise and its approach is in this sense superior to the more theoretical provenance of NSP and MDL. Nevertheless, we will show in section 7.4.2 that the physical number of spectrally distinct materials can not in general be regarded the best choice of model order estimate for the ABGP background subspace estimation algorithm. Finally we note that the most probable correlated noise source in this measurement configuration is comprised by speckle noise that is not completely eliminated by the speckle-reduction unit.

7.4.2 Background Endmember Extraction

In section 7.3.2 we showed that an underestimated model order strongly affects the detection performance for structured target detection algorithms. From this we conclude that the quality of the background endmember matrix is crucial for the resulting target detection performance. If a background spectrum is omitted (as is the case, if the model order is underestimated), target detection performance is expected to decrease. The endmember extraction method favored in the AMSD implementation in this work is the ABGP and it is therefore of interest to analyze its operation on measurements obtained by the hyperspectral image sensor.
We present the output of the ABGP algorithm for background endmember extraction in figure 7.14. The sample under test is a Polyamide substrate containing RDX contaminations.

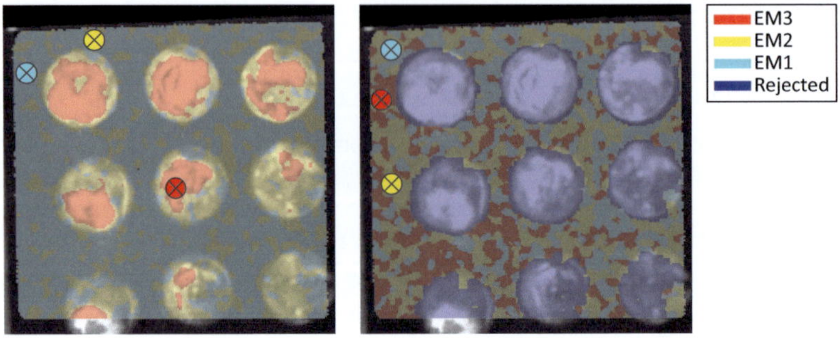

Figure 7.14: Endmember registration map of ABGP algorithm for background extraction. The measurement was performed on a Polyamide substrate with RDX contaminations from a distance of ≈ 1.4 m with a spectral range from 990 cm^{-1} to 1330 cm^{-1} and a spectral resolution of 1 cm^{-1}. Left: background registration map for ABGP with hypothesis PETN; right: same for RDX.

As in section 7.3.1, we show the endmember locations generated by the first part of the ABGP algorithm, together with the cluster membership of each pixel vector as a result of the stabilization stage. The endmember clustering results are given as color maps overlaid over a visible image of the sample. The ABGP outcome is given for the case of a target spectrum that is not contained in the observation data (i. e. PETN) and for the case of the contamination substance. A fixed value of $q = 3$ was chosen as total number of endmembers. In the case of non-existing target (i. e. the PETN hypothesis in this case) the ABGP chose the first two endmembers to account for the Polyamide substrate. The third endmember fell into an RDX contaminated area that is hence correctly considered background for this hypothesis. The clustering results show a random distribution of endmember one and two within the background area, indicating a fairly homogeneous substrate material and illumination. Under the RDX hypothesis, the spectra from the contaminated areas fall into the rejection class that is not used for background matrix generation. The three endmember spectra are selected by the ABGP algorithm at various locations in the scene that generate a spatially random clustering membership.

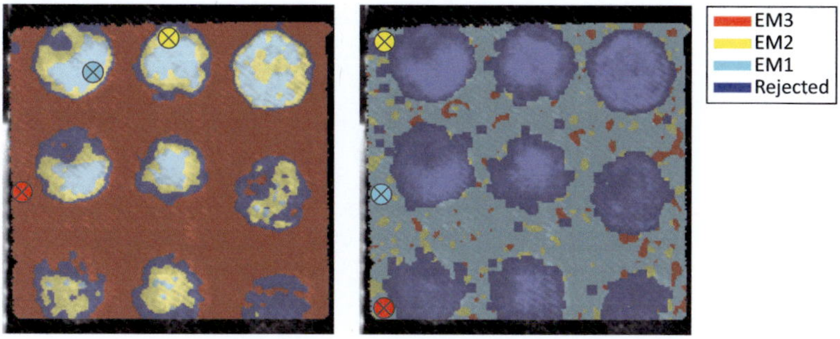

Figure 7.15: Endmember registration map of ABGP algorithm for background extraction. The measurement was performed on a jeans substrate containing RDX residues from a distance of $\approx 1.4\,\mathrm{m}$ with a spectral range from $990\,\mathrm{cm}^{-1}$ to $1330\,\mathrm{cm}^{-1}$ and a spectral resolution of $1\,\mathrm{cm}^{-1}$. Left: background registration map for ABGP with hypothesis TNT; right: same for hypothesis RDX.

Figure 7.15 shows the ABGP result for the case of a jeans cloth sample containing RDX contaminated areas. In this case, results for each of the two hypotheses of TNT and RDX contamination are given as colormaps overlaid over visible images of the sample. As for the previously discussed case of RDX contamination on Polyamide, the pixel vectors within the target contaminated areas fall into the rejection class. The three background endmember clusters are randomly distributed across the remaining scene. For the case of the TNT hypotheses, the first two endmembers generated by the ABGP come from areas contaminated by RDX, the first endmember being in a region of high contamination and the second at a fringe area containing a higher background contribution. It is only the third endmember that is chosen within uncontaminated background region.

A hint towards explanation of this behavior can be drawn from the spatial distribution of the normalized backscattering intensity, shown in figure 7.16. The mean backscattering signal intensity measured on the jeans cloth background is significantly lower compared to that observed from the contaminated areas.

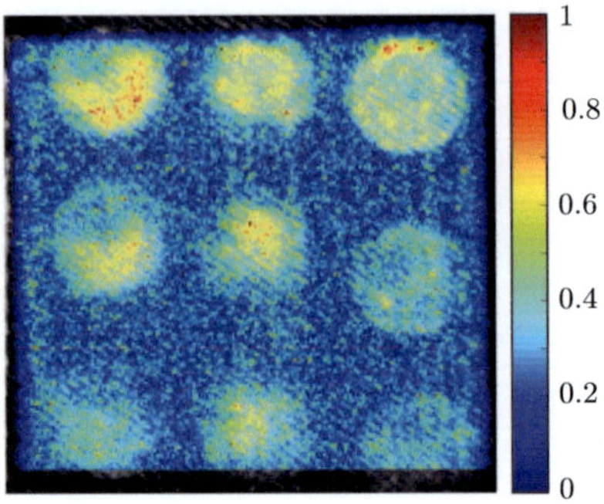

Figure 7.16: Spatial distribution of the normalized mean backscattering intensity of RDX on Jeans cloth. The mean energy in the background substrate backscattering signal is low compared to that observed from the target contaminated areas.

We recall at this point that the first part of the ABGP algorithm, is based upon a magnitude measure: in each iteration an orthogonal projection operator is employed to map the observed data into the subspace that is orthogonal to the space spanned by the background subspace generated so far. The method therefore intrinsically favors brighter areas in the image. Following these findings, one could be misled to choose an implementation of the ABGP that operates on a mean intensity-normalized representation of the hyperspectral image. The latter can be easily obtained by dividing each pixel vector by its mean intensity. This approach has the side effect of strongly amplifying not only the signal but also the noise of pixel vectors with low mean energy, especially at the edges of the illumination area. Particularly noisy spectra yield a higher magnitude in the orthogonal subspace and consequently this variant of the ABGP algorithm is drawn towards low energy pixels, which is for obvious reasons highly undesirable. We conclude that the stated property of favoring bright pixels is not so much a flaw, but can rather be regarded beneficial for the purpose. Yet, these findings emphasize the necessity of selecting a model order that is beyond the intuitive expected physical number of spectrally distinct materials in the image.

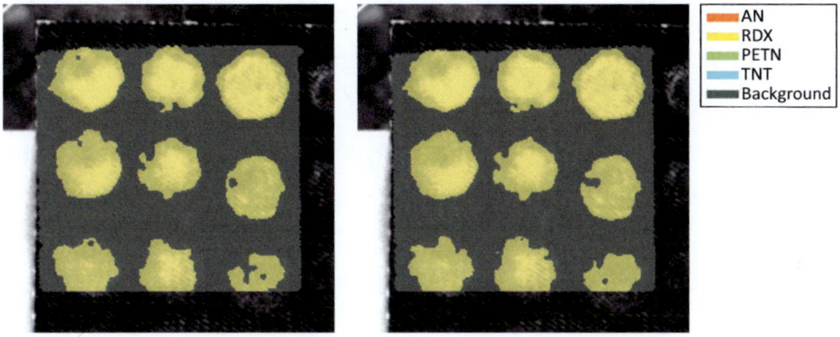

Figure 7.17: Detection of RDX residues on jeans cloth produced by the AMSD-ABGP algorithm with two different model orders. The detection image on the left was obtained using a model order estimate of $q = 5$ for the ABGP process. The MDL model order estimate $q = 71$ was used for the detection image on the right. The results suggest that overestimation of the model order has no adverse effect on target detection performance.

7.4.3 Impact of Model Order Mismatch

In section 7.4.1 we observed that the model order estimates generated by all considered estimation methods are subject to possible mismatches. While the PCA based method as well as the NSP and MDL approaches significantly overestimate the required model order, the SML algorithm yields a measure that is more closely related to the physical number of spectrally distinct materials in the scene. We demonstrated in the previous section that the latter might cause the ABGP to generate insufficient background representations. In section 7.3.2 we showed that the target detection accuracy of the ABGP based AMSD implementation is robust to overestimating the required model order. It is left to investigate the impact of model order overestimation on real hyperspectral image measurements. In figure 7.17 we present the target detection results obtained on an RDX contaminated jeans cloth sample using the ABGP based AMSD algorithm. The first result was obtained by operating the ABGP in a manual mode choosing $q = 5$ as model order estimate. The second detection result was generated, using the MDL estimate $q = 71$.

The two operation modes yield close to identical target detection results. This indicates that the required model order is significantly lower than that generated by the MDL estimation method. Despite significantly increased processing time, the ABGP algorithm ensures that target detection performance is however not degraded.

7.4.4 Short Distance Detection Results

Having analyzed the operation of the ABGP as background extraction method we now turn towards results obtained from the detection step on real-world measurements. Lead by conclusions following from section 7.3.6 we concentrate on the CFAR sub pixel target detection algorithm AMSD (in the implementation variant using the ABGP) and the ACE detector.

The results presented in the following were obtained from the same samples and measurement configuration as outlined at the beginning of the previous section 7.4.2, i. e. using the short range hyperspectral image sensor with a resolution of 1 cm^{-1} in the range of 990 cm^{-1} to 1310 cm^{-1}.

In figure 7.18 we present the detection results of the AMSD-ABGP target detection method on various combinations of jeans cloth and Polyamide substrates contaminated by traces of RDX, PETN and TNT. We observe that all RDX contaminated areas on the Polyamide substrate successfully generated positive detection results with only minor false alarms raised for the competing substances (PETN, TNT an AN).

For the case of the PETN contaminated Polyamide sample, the system also identified target pixels within all PETN regions, again raising only minor false alarms for the competing substances. For the case of TNT contaminants on Polyamide the detection outcome is slightly inferior, as one of the contamination areas failed to cause an alarm. All other contaminated areas were successfully identified and only few false alarms are obtained.

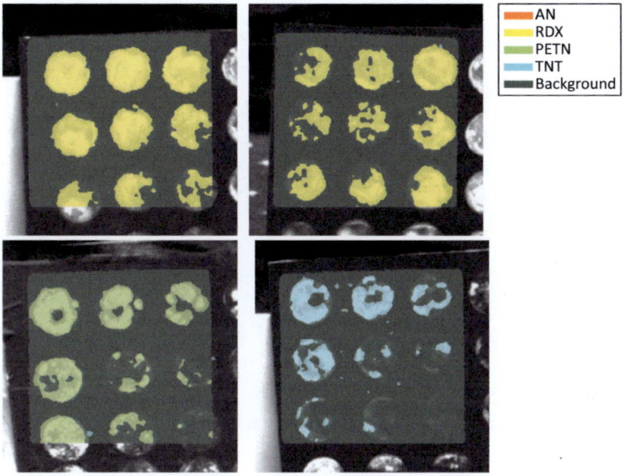

Figure 7.18: Detection results obtained using the AMSD-ABGP target detection algorithm. Top: Detection of RDX contaminated areas on a Polyamide substrate. Top left: high RDX concentrations, top right: lower concentrations. Bottom: detection of PETN (left) and TNT (right) contamination areas on a Polyamide substrate.

These results are well comparable to the outcome of the simulations shown in section 7.3.7, where we observed that RDX is expected to yield slightly better performance than PETN, TNT and AN in descending order.
Having the detection results given in figure 7.18 it is inviting to employ the classification results to reconfirm applicability of the LMM for the proposed measurement method. As in section 6.5, we employ the LMM to attempt to unmix the observed spectra from the areas that were classified as contaminated by the AMSD-ABGP algorithm. If the LMM holds, linear unmixing is expected to be capable of extracting the library spectrum from the contaminated mean spectrum, given the background spectrum and the appropriate library spectrum. In figure 7.19 we present results from spectral unmixing, for the case of the detection results acquired on the TNT contaminated Polyamide sample.

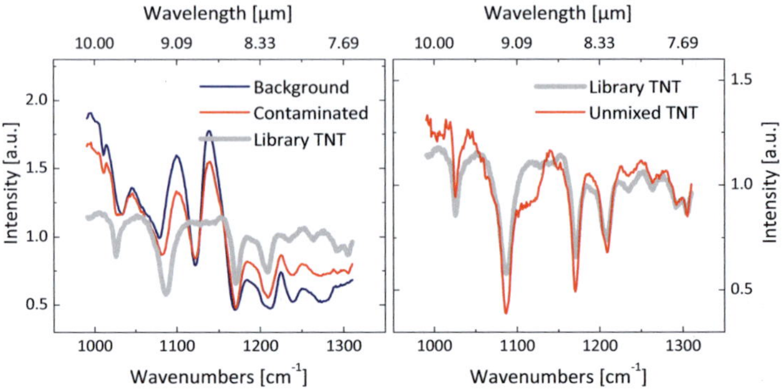

Figure 7.19: Validation of linear mixture model for the case of TNT on Polyamide substrate. Using the background and contaminated spectra obtained by spatial averaging over the corresponding regions as generated by the detection step we obtain a close match of the TNT library spectrum.

It is well observable that the mean spectra of background and contaminated area are highly similar. Nevertheless, using the linear mixture model and the spectral unmixing process outlined in section 6.5 the TNT library spectrum can be almost perfectly gained from the background and contaminated spectrum with a low residual norm.

7.4.5 Comparison of Detection Algorithms

Besides the AMSD target detection algorithm we considered target detection performance of a variety of other approaches in section 7.3.6: the ACE detector as unstructured detection algorithm, the OSP detector as structured target detector and the MF, CEM and NCC algorithms as representatives of full pixel target detection methods. In figure 7.20 we present a comparison of the detection results obtained by the mentioned methods for the case of PETN residues on a Polyamide substrate. From these results we conclude that the performance differences among the model based sub pixel detectors (AMSD-ABGP, ACE and OSP) are marginal and significantly outperform the full pixel detectors.

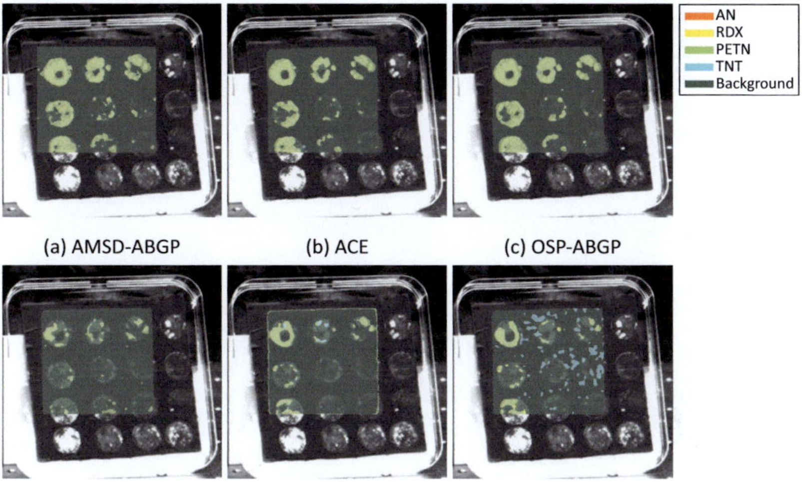

(a) AMSD-ABGP (b) ACE (c) OSP-ABGP

(d) CEM (e) MF (f) NCC

Figure 7.20: Comparison of detection results obtained by the considered sub pixel detection algorithms (top row) and full pixel detectors (bottom row) of PETN residues on a Polyamide sample.

While the sub pixel detectors are still capable of detecting low residue concentrations, the full pixel detectors yield successful detection results only for areas with high target coverage. Among the full pixel detectors, the CEM detector slightly outperforms the MF and NCC detector. The MF is both prone to false alarms of TNT as well as PETN false alarms at the fringes of the illuminated area. High false alarm rates for TNT are observed in the detection map generated by the NCC detector. These observations match well to the simulation results in section 7.3.6.

Whereas – both in artificial and real-world hyperspectral image data – only marginal performance differences were observed among the structured detection algorithms, only the AMSD and the ACE detector have the CFAR property which is beneficial under various conditions (section 7.3.5). Several detection results using the AMSD-ABGP algorithm were already shown.

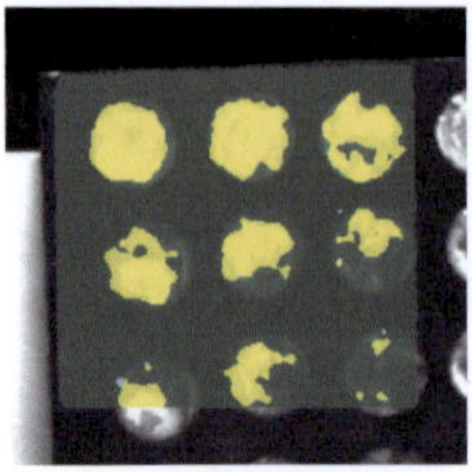

(a) RDX on Polyamide

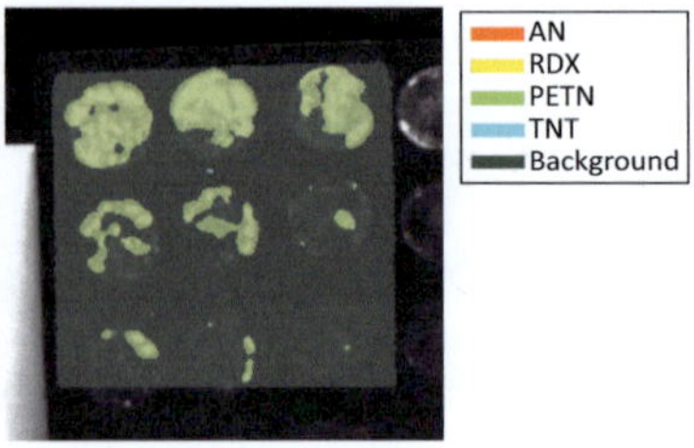

(b) PETN on Polyamide

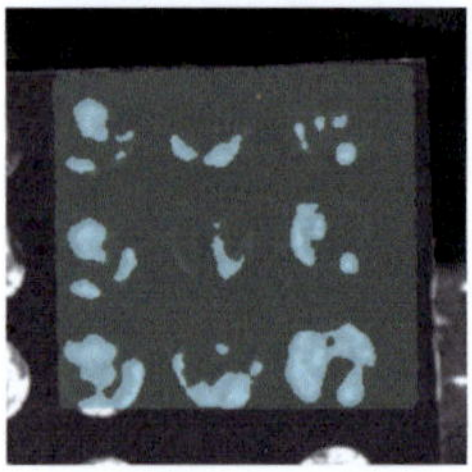

(c) TNT on Polyamide

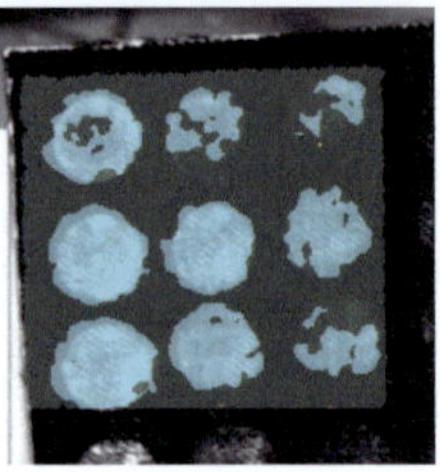

(d) TNT on jeans

Figure 7.21: Detection of RDX, PETN and TNT residues using the ACE target detection algorithm on various samples.

In the following, we present detection results obtained on real measurement data using the ACE target detection algorithm in figure 7.21. As expected, the ACE detection results are well comparable to those obtained from the AMSD-ABGP algorithm. All RDX and PETN contaminated areas on the Polyamide substrate generated positive detection results. The TNT contaminations on Polyamide and jeans cloth were also successfully identified with no false alarms raised for the competing substances.

We conclude that most results obtained from experiments on artificial hyperspectral image data are well transferable to analysis of real measurement data. Finally, both considered sub pixel detection algorithms suggest to comprise promising approaches for substance detection in hyperspectral image data acquired using the proposed measurement method.

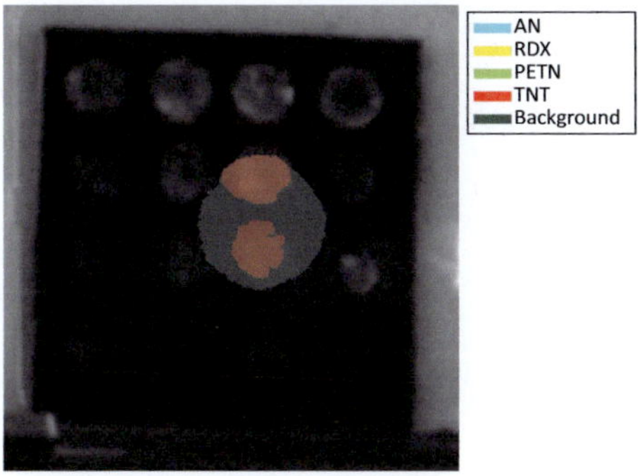

Figure 7.22: Detection of TNT residues on jeans cloth from a distance of 15 m using the long range setup of the hyperspectral imaging sensor. The detection results were obtained using the AMSD-ABGP detection algorithm.

7.4.6 Long Range Detection Results

In a final step we now turn towards evaluation of hyperspectral image data, acquired using the long range hyperspectral image sensor presented in section 6.1. To enable long range measurements, this setup differs from the short range hyperspectral image sensor predominantly in terms of collection and projection optics. As stated earlier, customized Schmidt-Cassegrain mirror telescope with a focal length of 1 m and an f-number of $f\,/\,3$ serves as collection optics that provides sufficient spatial resolution for employing the suggested target detection algorithms for detection of substance traces across longer distances. The projection optics were chosen as to achieve a circular top-hat beam profile with a diameter of about 5 cm at a typical operating distance of ≈ 15 m. In figure 7.22 we present detection of TNT residues on a jeans cloth substrate using the long range hyperspectral image sensor. The measurement was performed over the full tuning range from 1000 cm^{-1} to 1300 cm^{-1} at a resolution of 1 cm^{-1} and a distance of ≈ 15 m. The AMSD-ABGP detection algorithm was employed for target detection. The contaminated areas were successfully identified and no false alarms were generated for the competing substances.

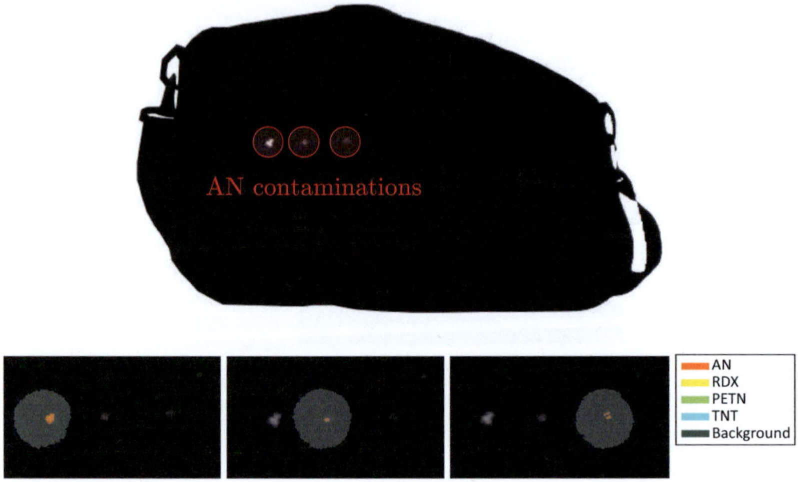

Figure 7.23: Top: Polyamide carrying bag containing several Ammonium Nitrate contaminations. Bottom: detection results obtained using the long range hyperspectral image sensor from a distance of $\approx$17 m. All AN contamination areas were successfully detected by the AMSD-ABGP detection algorithm.

In figure 7.23 we present measurement results using the same setup for detection of traces of Ammoniumnitrate (AN) on the surface of a Polyamide carrying bag from a distance of $\approx$17 m. Again, the detection results were obtained using the AMSD-ABGP detection algorithm. Three different AN contamination areas were successfully identified with no false alarms generated for the competing substances.

An analysis of the corresponding spectra is given in figure 7.24. In comparison to the spectra obtained by the short range hyperspectral image sensor the acquired signals have a considerably lower SNR. This is caused not only by the larger measurement distance, but also by degraded imaging quality (section 6.1) and atmospheric scintillation effects. In addition, the AN spectrum has only few characteristic spectral features in the considered wavelength range at around 1042 cm^{-1}.

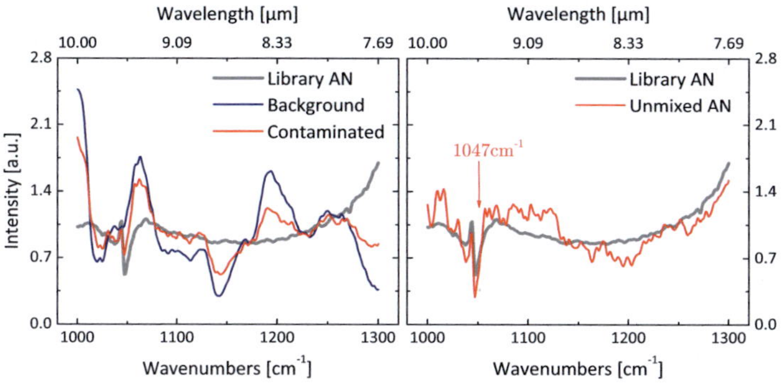

Figure 7.24: Spectral analysis of the AMSD-ABGP detection results on long distance measurement of AN on Polyamide. Left: the mean spectrum of the contaminated area shows the AN characteristic spectral feature at 1047 cm^{-1}. Right: linear un-mixing allows to obtain the target spectrum from background and contaminated spectrum.

The corresponding spectral feature is observed in the mean spectrum of the contaminated area in figure 7.24 that otherwise predominantly mimics the background spectrum. Nevertheless, using linear unmixing, the library spectrum can be obtained from the contaminated and background spectra using the linear mixture model fairly well.

7.5 Summary and Conclusion

In this chapter we presented a detailed analysis of the data analysis methods introduced in chapters 3 and 4 under special consideration of artifacts and general properties specific to the proposed measurement methods. Results obtained on artificial hyperspectral data served as guides for assembly of a data processing and analysis framework operational on real hyperspectral image data. We presented target detection results for various substance traces using both the short range and the long range variant of the hyperspectral image sensor. Both the ACE detector as well as the AMSD target detection algorithm implemented with the proposed ABGP background extraction method showed well applicable for this task. Compared to the competing methods, the proposed MDL model selection method proved superior for analysis of artificial data, yet highly sensitive to correlated noise under which it selects significantly overestimated models. Nevertheless, it was shown that – while highly sensitive to overly simple models – the AMSD-ABGP target detection process is robust against overestimated model orders.

8 Extension to Real-Time Spectroscopic Sensing

In the previous chapter we provided a detailed analysis of the considered data analysis algorithms applied to hyperspectral images obtained by the proposed tunable long-wave infrared (LWIR) laser based backscattering spectroscopy principle. The hyperspectral image acquisition process is however time consuming. The most relevant factors in this regard are the process of mechanical wavelength tuning and the limited sensor frame rate, in combination with online preprocessing. In a typical sensor configuration, a full hyperspectral image with a $1\,\mathrm{cm}^{-1}$ resolution in the range from $1000\,\mathrm{cm}^{-1}$ to $1300\,\mathrm{cm}^{-1}$ is acquired and ready for analysis in $\approx 20\,\mathrm{s}$. Due to the measurement principle, any relative movement of target or sensor render the resulting measurement data essentially unusable. Thus, application of this measurement technology is currently reliant on fairly cooperative or static scenes. In this section we will present a modification of the measurement principle that allows real-time spectroscopy and consequently enables remote substance detection in dynamic environments at the expence of imaging capability in a stricter sense. This is achieved by replacing the conventional mechanically tuned External Cavity Quantum Cascade Laser (EC-QCL) by a rapid wavelength-scanning Micro Opto Electromechanical System (MOEMS) tuned EC-QCL that was developed in collaboration with Fraunhofer IPMS [24, 47]. The technique allows a full spectral scan at a time-scale of $\approx 1\,\mathrm{ms}$ rendering measured spectra invariant to relative motion of sensor or sample that might be expected in relevant scenes.

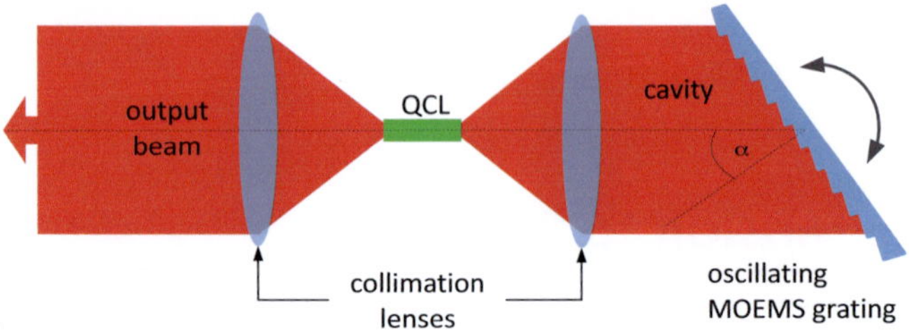

Figure 8.1: The MOEMS EC-QCL uses a MOEMS grating for wavelength scanning, that oscillates at it's resonance frequency. Hence, emission wavelength is a function of time.

The final measurement system employs a single element detector to produce a spectral acquisition rate of around 1 kHz at an average spectral resolution of approximately 1.5 cm^{-1}. We begin with a description of the illumination source in section 8.1. In section 8.2 we present the experimental setup and follow with explanation of the data acquisition process in section 8.2.1. We close with a presentation of measurement results in section 8.3 and an analysis of real-time operation of the data analysis process.

8.1 The MOEMS EC-QCL Illumination Source

In this section we introduce the MOEMS External Cavity Quantum Cascade Laser (MOEMS EC-QCL) as rapid wavelength scanning variant of the EC-QCL presented in section 6.1.1. The latter will be referred to, using the term step scanning EC-QCL in the following. As outlined in section 6.1.1 the step scanning EC-QCL comprises an optical grating as wavelength selective element that is mounted upon a conventional piezo-driven rotation stage. The latter allows control of the emission wavelength by manipulating the angle of incidence. We show an illustration of the fast scanning MOEMS EC-QCL principle in figure 8.1. As in the layout of the static EC-QCL, the QCL chip is operated using an external cavity resonator in a Littrow configuration. In this variant a grating is employed as wavelength selective feedback element that is etched into the surface of a MOEMS mirror.

Figure 8.2: MOEMS EC-QCL light source that was developed in collaboration with Fraunhofer IPMS. Left: compact design allows integration into measurement systems with a small footprint. Right: close-up of laser chip with collimation lenses in external cavity resonator and MOEMS grating.

In figure 8.2 we show a realization of a MOEMS EC-QCL. The silicon based MOEMS scanning grating was designed for a resonance frequency of 1 kHz with a maximum deflection amplitude of 10°. The surface has a diameter of 5 mm and has a 133 mm^{-1} groove density diffraction grating, optimized for maximum reflectivity into the desired diffraction order, etched into it. The electrostatically driven scanning plate is suspended by torsional springs and it's angular position is continuously measured by an integrated position sensor realized by four piezo sensors that are connected to a Wheatstone bridge. The trajectory can be continuously monitored using the position measurement. In addition, the sensor unit produces a trigger signal when the grating plate crosses the zero angle position. This signal is used for wavelength synchronization as described in section 8.2 in more detail. The oscillatory trajectory of the grating plate renders the grating angle $\alpha(t)$ a function of time and causes the laser chip's emission wavelength to be altered accordingly. The trajectory is obtained as sum of the constant offset angle α_0 and the time dependent oscillation $\Delta_\alpha(t)$. The latter is sinusoidal in time with the MOEMS resonance frequency f_r and amplitude A_0, giving:

$$\alpha(t) = \alpha_0 + \Delta_\alpha(t) = \alpha_0 + A_0 \cdot \sin(2\pi f_r t + \phi_0). \tag{8.1}$$

The constant phase correction offset ϕ_0 in equation 8.1 is a design parameter that allows to account for a non-zero phase difference between the trigger signal and the true equilibrium position in the analysis of measurement data. Using these results and the grating function (equation 2.3) we finally obtain the time dependent emission wavelength function $\lambda(t)$ as

$$\lambda(t) = 2g\sin(\alpha_0 + A_0 \cdot \sin(2\pi f_r t + \phi_0)). \tag{8.2}$$

As in the imaging spectroscopy setup, we operate the QCLs in a low duty cycle pulsed mode. Consequently, the emission wavelength function of the MOEMS EC-QCL is not a continuous function $\lambda(t)$, but rather a discrete time series $\lambda[t]$. The latter is obtained from the continuous representation by employing the laser pulse as sampling signal. Hence, the parameters of the laser pulse train define the spectral resolution of the resulting measurement device. An analysis of the latter will be subject of section 8.2.2.

8.2 Experimental Setup

The principle experimental setup for the real-time spectroscopy device is well comparable to the hyperspectral imaging setup, presented in chapter 6. A tunable laser source – in this case the MOEMS EC-QCL – is used to illuminate the surface to be analyzed and the diffusely backscattered light is collected by a single element signal detector. The latter replaces the image sensor in chapter 6. A BaF_2 beam-splitter is used to divert a portion of the laser output beam into a gold coated integrating sphere, to serve as reference signal. This measurement principle is illustrated in figure 8.3
In the experimental setup used for data acquisition in this work, TE cooled photo-voltaic Mercury Cadmium Telluride (MCT) detectors (VIGO Systems PVMI-2TE-10.6) were used as signal and reference detectors. The MOEMS EC-QCL was equipped with a heterocascading QCL chip tunable in the range from 1080 cm^{-1} to 1370 cm^{-1}.

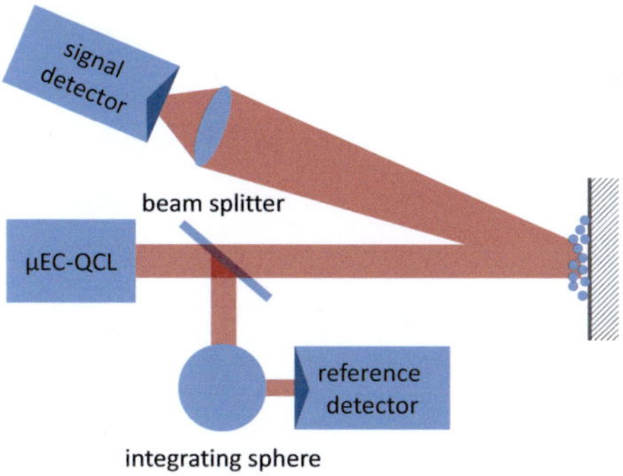

Figure 8.3: The MOEMS EC-QCL (here denoted µEC-QCL) is used to illuminate the scene and the diffusely backscattered light is measured by a fast single element photo-voltaic MCT detector. An identical detector serves as reference detector.

The MOEMS grating (133 mm^{-1} groove density) was oscillating slightly above its resonance frequency (975.6 Hz) at an amplitude of $\approx 5°$. The offset α_0 was set to $\approx 33.6°$, to roughly match the center of the QCL chip's gain spectrum at $\approx 1200\,\text{cm}^{-1}$. The laser pulse frequency was set to 390.24 kHz to obtain a burst of 400 pulses per MOEMS oscillation period. The laser pulse width was set to 200 ns giving an effective duty cycle of 8.0 %.

8.2.1 Fast Data Acquisition

In this section we outline the real-time data acquisition workflow for real-time spectroscopic measurements using the MOEMS EC-QCL and follow with an analysis of the achievable and expected spectral resolution under the given operation conditions.

The rapid wavelength tuning in combination with the low duty cycle measurement requires a fast data acquisition setup for reliable spectroscopic measurements. In this work a digital lock-in amplifier (Zurich Instruments UHFLI) with integrated boxcar averaging capability was employed for this task. The lock-in amplifier is equipped with two independent lock-in units that – given a constant reference frequency – allows phase-dependent positioning of a boxcar averaging windows for each of two analog input channels. The laser pulse trigger is fed into the lock-in amplifier as reference frequency for the harmonic oscillators, and is employed for correct positioning of the digital boxcar averaging windows. The latter are used to average the analog input signals provided by signal and reference detectors, during active emission of the QCL chip, omitting the noise signal observed during the laser-off phase. This approach requires a constant frequency laser pulse signal to enable the lock-in technique. However, for reproducible spectral selective measurements, a strict phase coupling between laser pulse and MOEMS oscillation is required. The phase coupling ensures that the laser pulses occur at defined and reproducible oscillation phases in every oscillation period. This synchronization is achieved by operating the laser pulse generator in a burst mode. The MOEMS zero-phase synchronization signal is fed into the laser pulse generator and issues a burst of k pulses at a frequency $k \cdot f_r$ where f_r denotes the MOEMS oscillation frequency. The number of pulses per oscillation period is in our setup chosen to read $k = 400$ and determines the limiting spectral resolution of the measurement system as will be subject to a more detailed discussion in the following section.

The approach mentioned so far yields two boxcar averaged signals: signal and reference detector values that are obtained from time equidistant laser pulses, which were simultaneously generated at fixed positions of the MOEMS oscillator. To obtain a laser independent signal and correct for wavelength dependent atmospheric transmission effects, the boxcar averaged signal value is divided by the boxcar-averaged reference value. We then employ the MOEMS synchronization signal to organize the data chunks to the analysis computer in a defined manner. I. e. a $k \times l$ matrix is generated that contains the k time averaged reference signal values as entries row-wise. Row switching is induced by the MOEMS synchronization signal. This ensures that the l values in a column of the resulting measurement matrix correspond to the same phase offset of the MOEMS oscillation. The matrix is transferred to the analysis computer for further analysis.

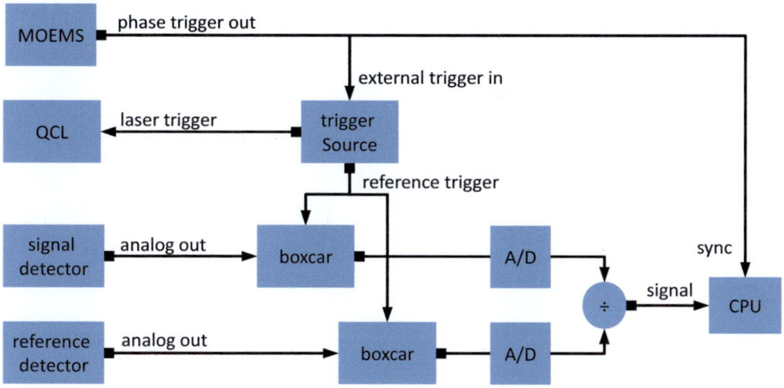

Figure 8.4: Fast signal acquisition for real-time spectroscopic sensing using the MOEMS EC-QCL light engine. The MOEMS zero-phase signal is used to synchronize the laser pulse trigger signal to the MOEMS oscillation phase. Single pulses are extracted from signal and reference detector using boxcar averaging. The resulting referenced measurement signal is obtained by dividing the boxcar-averaged signals. Synchronization to the MOEMS phase signal is required for organizing the data chunks for further spectroscopic analysis.

The effect of the boxcar averaging technique is shown in figure 8.5. For this simulation, a resonance frequency of 1 kHz with an amplitude of 4.5° and an angular offset of 33.6° (grating constant 133 mm^{-1}) was chosen. We show the trajectory of the MOEMS mirror within the first 20 µs of an oscillation period, after crossing the equilibrium position (where the trajectory's rate of change is highest). The equidistant laser pulses (width: 200 ns) are surrounded by the box-car window that – with a width of 210 ns – was chosen slightly larger than the laser pulse to compensate for potential frequency or phase jitter. Whereas the designated laser emission signal caused by the MOEMS' oscillatory motion changes over several wavenumbers, the boxcar output signal remains constant between two laser pulses. This allows us to use the laser repetition frequency itself as sampling frequency of the boxcar averaged signal, which is significantly beyond that required to resolve the narrow laser pulses in the original low duty cycle signal.

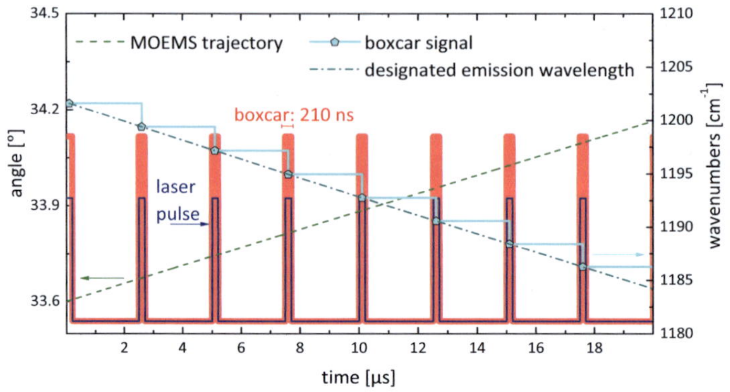

Figure 8.5: Boxcar averaging for low duty cycle measurement using the MOEMS EC-QCL as illumination source

We note that as a side-effect this measurement principle generates a discrete sampling of the backscattering spectrum with a sampling rate constant in time, rather than in wavenumbers as was the case with the hyperspectral imaging setup presented in chapter 6. Due to the nonlinear motion of the MOEMS grating, this yields a non equidistant sampling in the wavenumber domain. Before we analyze the consequences of this in the following section, we will turn to another effect observable in figure 8.5: the MOEMS position changes between the beginning and the end of a single laser pulse. The question may arise, if this changes the emission wavelength, and thus introduces spectral blurring into the resulting spectra.

The effect of the angular motion on emission wavelength within a single pulse under the given operation conditions is shown in figure 8.6. We show the angular trajectory of the MOEMS and the designated (i. e. mathematical) emission wavenumber difference, covered during the laser pulses at various phases of the MOEMS oscillation period. We refer to the designated emission wavenumber covered within a constant time frame as *in-pulse resolution* and note that – following the oscillation's angular rate of change – it is a function of the MOEMS phase.

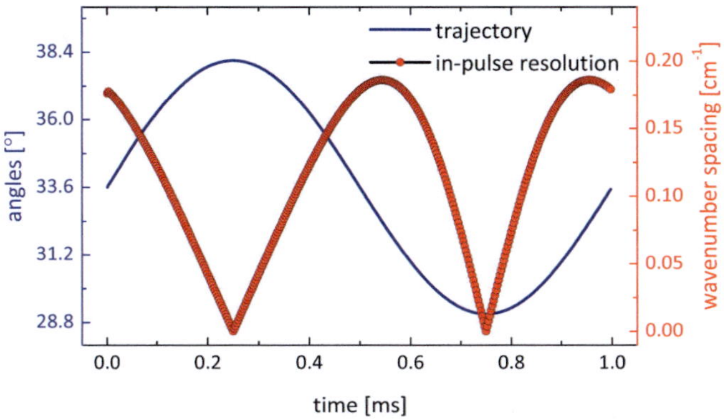

Figure 8.6: The phase dependent angular rate of change during a laser pulse causes the designated emission wavenumber to shift. This limits the theoretical resolution limit under the given parameters to 0.19 cm^{-1} which resides well below the pulsed EC-QCLs physical emission bandwidth of $\approx$ 1 cm^{-1}.

We observe that the maximum wavenumber spacing (i. e. lowest spectral resolution) is obtained at the zero crossings of the MOEMS oscillation, where the angular rate of change is largest. At these phase offsets the in-pulse resolution approaches 0.19 cm^{-1}, which however is well below the QCL's physical emission bandwidth. The latter is expected to be around 1 cm^{-1} in the given external cavity resonator configuration under pulsed operation (see section 2.2). We therefore expect the emission wavelength to be unaffected by the MOEMS motion during the pulses and consequently the backscattering spectra obtained using the MOEMS EC-QCL to match those, obtained by the static EC-QCL in the imaging spectroscopy setup.

8.2.2 Spectral Resolution

In the same way as the in-pulse resolution considered in the last section is a function of the MOEMS oscillation phase, the equidistant laser pulse frequency yields a non constant spectral sampling in the wavenumber domain.

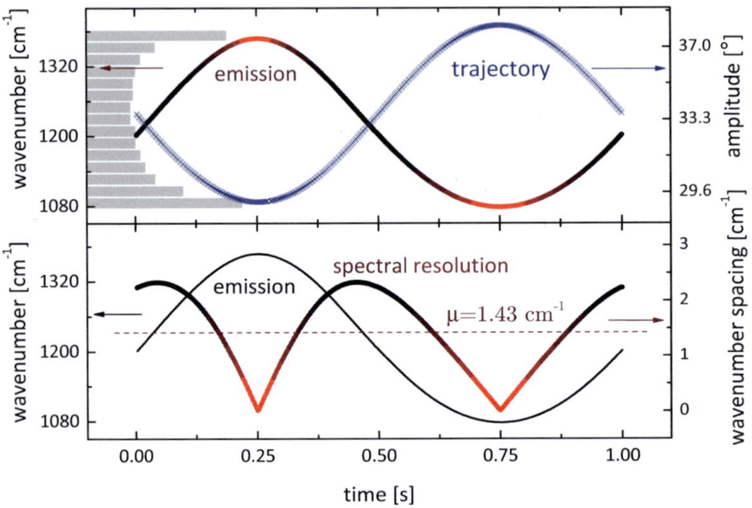

Figure 8.7: Spectral resolution obtained in the real-time measurement configuration. The sampling density is color coded in the wavenumber emission diagram with red indicating higher resolution (i. e. lower wavenumber rate of change).

In section 7.3.3 we showed that spectral resolution has direct impact on the resulting detection performance: the detection performance was significantly reduced for spectral resolutions beyond 2 cm^{-1}. In this section, we give an analysis of the spectral resolution obtained by the MOEMS EC-QCL measurement setup in the following section.

In figure 8.7 we present a simulation of the expected spectral resolution obtained when using the MOEMS EC-QCL as spectrally selective illumination source. Given a 133 mm^{-1} grating and an equilibrium angle position of 33.6° we obtain a zero-amplitude emission of 1202 cm^{-1}. The maximum MOEMS amplitude was set to 4.5° and an oscillation frequency of 1 kHz was chosen for the MOEMS element. The latter in combination with a laser pulse frequency of 400 kHz yields exactly 400 laser pulses (i. e. spectral elements) evenly distributed across an MOEMS oscillation period.

The angular trajectory of the MOEMS element is given as scatter plot showing the amplitude at the equidistant time-stamps that are defined by the laser pulse train. Considering the distribution of associated wavenumbers emitted by the MOEMS EC-QCL at these time stamps, we observe that the long- and short-wave cutoff wavelengths of the emission spectrum are over-represented in the emission histogram. This is caused by the non equidistant sampling in the wavenumber domain which renders the spectral wavelength resolution dependent upon the MOEMS oscillation phase. Whereas a mathematical resolution of 0.007 cm^{-1} is achieved at the turning points of the oscillatory motion, it decreases to a resolution of 2.3 cm^{-1} when passing the equilibrium position. The cumulative spectral resolution over the full oscillation period amounts to 1.43 cm^{-1}.

In addition, the emission histogram as well as the resolution graph suggest that the sampling resolution in the wavenumber domain is skewed towards the long wave fringe of the spectrum. On average, more long wave pulses are generated than short wave pulses. This is caused by the MOEMS oscillator's non central equilibrium position: a fixed angle delta generates a decreasing wavenumber rate of change for increasing angles.

We note that – if the relevant parameters are well matched to the QCL emission spectrum – the non constant spectral sampling effect is beneficial for the proposed measurement technique as it intrinsically compensates the laser chip's emission intensity distribution. The latter decreases towards the long- and short-wave fringes of the emission spectrum. This is compensated by the fact that the laser's emission linewidth is significantly over sampled. Likewise, the weaker spectral resolution observed when passing the equilibrium position, usually matches the peak power wavelength in a well configured setup. In addition, the marginal resolution of 2.3 cm^{-1} is still significantly beyond typical spectral features observed in spectroscopy of solids (e. g. section 6.4).

8.3 Measurement Results

In this section we will present experimental results obtained with the rapid wavelength scanning spectroscopy measurement scheme. The measurements were obtained using the setup as described above at a distance of ≈ 1 m. The collimated output beam of the MOEMS EC-QCL is used directly for target illumination without any further beam shaping. This leads to a Gaussian beam profile having a 4σ diameter of about 1 cm at the chosen measurement distance. A large gold coated plain mirror is used to deflect the illumination beam onto the surface of the optical table, allowing for a constant measurement distance and focus. A green pilot laser is used to indicate the measurement spot. The backscattered radiation is focused onto the detector located next to the illumination source using a 2" ZnSe lens.
In figure 8.8 we show a comparison of the spectroscopy results obtained with the imaging spectroscopy setup and the rapid wavelength scanning backscattering spectroscopy principle [30]. For the results shown, 50 consecutive backscattering spectra of PETN and TNT substance residues were averaged yielding an effective acquisition time of ≈ 51 ms. The substance residues were deposited on a weakly scattering substrate (painted metal plate) to avoid background spectral contribution. The comparison spectra were taken from the substance library introduced in section 6.4.
The spectral measurement range of the rapid wavelength scanning approach is achieved using a single QCL as opposed to the dual core solution operated in the image backscattering spectroscopy setup. Consequently, the spectral measurement range of the latter is considerably larger. However, the obtained substance spectra match well in the overlapping region. Whereas these results are promising and target detection can be easily achieved using any full pixel detection method, it is clear that background contribution is to be expected in a more general measurement scenario. In this case, a sub pixel algorithm is required for reliable detection, which however requires knowledge of the background spectral characteristics.

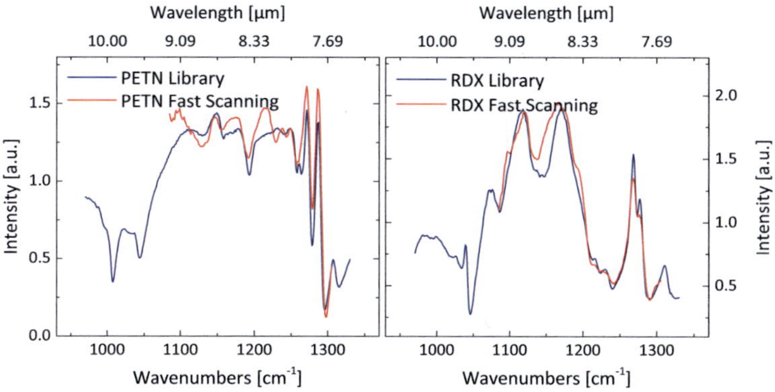

Figure 8.8: Real-time spectroscopy measurements of pure PETN and RDX in comparison to the library spectra that were measured with the imaging backscattering spectroscopy setup. The substance residues were deposited on a weakly scattering surface (painted car plate) and measured in ≈ 51 ms.

For the case of data acquired with the hyperspectral image sensor, this information was extracted from the available image data using the Adaptive Background Generation Process (ABGP) background extraction algorithm. A crucial boundary condition for applicability of the latter is however that pure background spectra are existent in the hyperspectral data to be analyzed. This can be achieved using the real-time spectroscopic measurement setup, by spatial variation of the measurement spot – either by variation of the pointing direction of the measurement device or by moving the sample while keeping the measurement spot constant. To simulate the former in a first evaluation of the principle, we employ the second approach in the following and present measurement results acquired on a PETN-contaminated Polyamide substrate sample. An approximation of the measurement path over the sample is shown in figure 8.9. The movement of the sample was performed manually at a non constant speed and the full motion was performed in about 1 s. A constant stream of spectra was measured within this time using the fast spectroscopy principle.

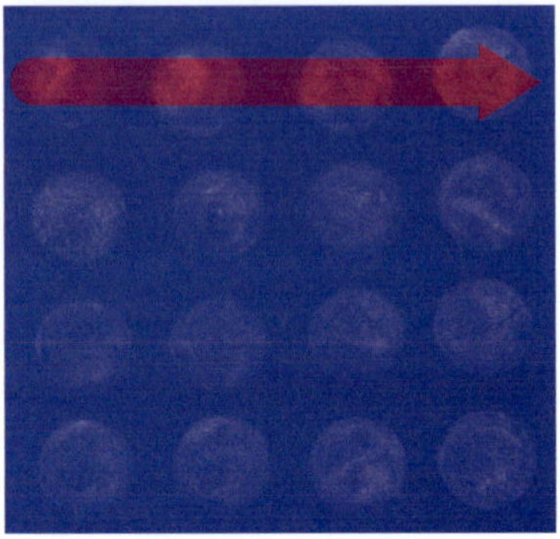

Figure 8.9: Approximate measurement path on the RDX contaminated Polyamide sample. The sample was moved at a non constant speed under the fixed measurement spot, while the backscattering spectra were continuously recorded using the fast spectroscopy setup.

We present a color-coded backscattering intensity map of the hyperspectral measurement data obtained by this experiment in figure 8.10. The Adaptive Matched Subspace Detector (AMSD) with ABGP background extraction algorithm was employed for target detection, to check each acquired spectrum (that correspond to pixel vectors in the imaging variant of the spectroscopy device) for contamination with either one of the hypotheses PETN, RDX, TNT and AN. The detection result is given in the color coded ribbon at the top of figure 8.10 (top). Several positive detection results for PETN were generated and no false alarms were raised for any of the competing spectra in the hypotheses set.
The hyperspectral measurement data obtained on a similar measurement of RDX samples on a Polyamide substrate are shown in figure 8.10 (bottom), which also contains the output of the AMSD algorithm with ABGP background extraction in the ribbon at the top. As for the case of the PETN traces on Polyamide, several positive detection results were generated and no false alarms were obtained for the competing substances.

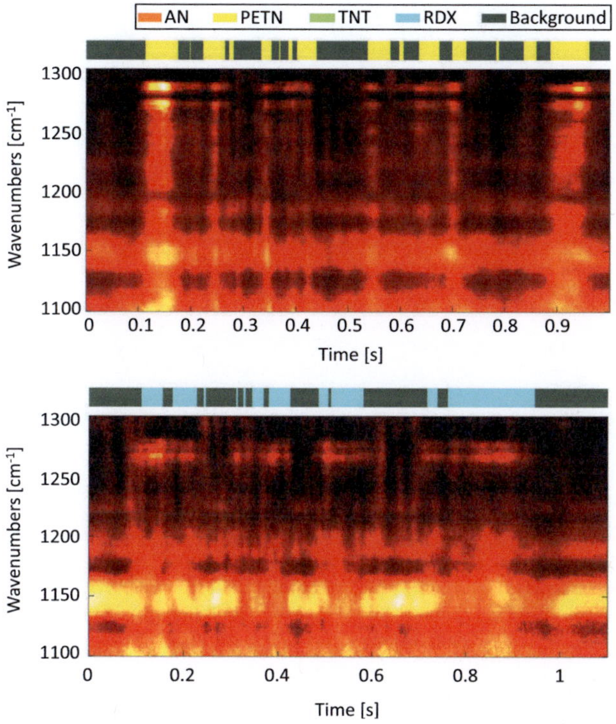

Figure 8.10: Detection of PETN (top) and RDX (bottom) contaminations on a Polyamide surface using real-time backscattering spectroscopy. The backscattering intensities are color-coded. The detection results were obtained by the AMSD-ABGP algorithm.

In figure 8.11 we present an analysis of the classification results, obtained on the PETN contaminated Polyamide sample. We show the mean spectrum obtained by averaging over the pixel vectors classified as *background* in comparison to the pixel vectors classified as *PETN-contaminated* and the pure PETN library spectrum. The characteristic peaks of the PETN library spectrum are well observed in the contaminated average spectrum, and only weakly in the background mean spectrum.

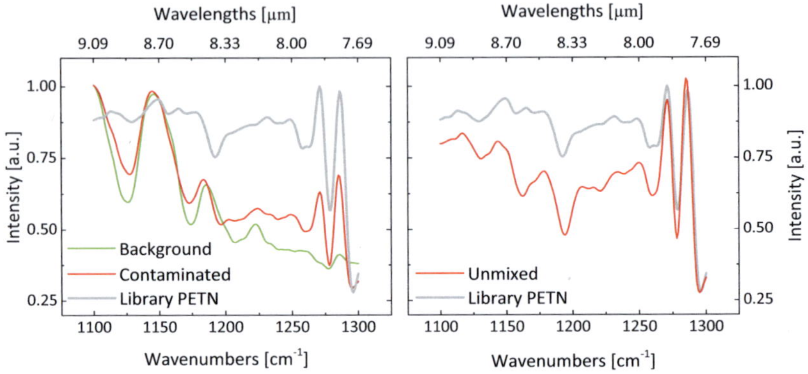

Figure 8.11: Analysis of background and PETN contaminated region spectra, obtained using the output of the AMSD-ABGP detection algorithm.

This indicates that the algorithm successfully discriminated background and PETN contaminated spectra in the observation set. We also show the result of spectral unmixing as suggested in section 6.5: the contaminated mean spectrum can be successfully unmixed to match the library spectrum using the background mean spectrum and the Linear Mixture Model (LMM). The latter indicates that the LMM is a valid mixture model for the measurement data produced by the proposed fast scanning spectroscopy principle.

8.3.1 Empirical Validation of Real-Time Data Processing

Whereas the data acquisition process outlined in section 8.2.1 allows for fast measurement acquisition and thus for real-time detection, this might not be necessarily true for the data analysis process. The AMSD target detection algorithm that was favored for analysis of the measurement data generated by the hyperspectral image sensor requires a background endmember extraction algorithm for operation. The ABGP algorithm was suggested for this task that comprises the Adaptive Target Generation Process (ATGP) algorithm in combination with a stabilization step based on data

clustering (section 3.5.2). In the following we will show, that the proposed data processing methods are in general capable of real-time operation by measurement of the computation time required for these steps.
As stated before, for applicability of sub pixel target detection algorithms like the AMSD, a hyperspectral data set is required that contains pure background information. This requires operating the measurement device in a spatial scanning mode. We assume that the data containing background-only spectra is accumulated within a data set recorded in a specific time frame of length Δt. It is therefore sufficient, to estimate the time required to analyze a hyperspectral data chunk recorded within this time that for simplicity we choose to be $\Delta t = 1\,\mathrm{s}$.
We note that every substance of interest in the library requires a full detection run, i. e. the cumulative processing time is expected to scale linearly by the number of target substance hypothesis in the library (see section 3.7 for more details). The processing times stated in the following were measured on a 64 bit Windows 7 operated PC with an Intel® Core™ i5-4590 CPU running four cores at 3.3 GHz (3.7 GHz) and equipped with 8 GB RAM. Data analysis was implemented and executed in Matlab (R2016b) without utilization of parallel processing or GPU support.
The RDX contaminated Polyamide sample measurement presented in section 8.3 served as evaluation set, containing a total of 1.1 s of data and i. e. a total number of 1007 spectra. The average processing time of the ABGP background extraction process was measured to be 64 ms per time frame and target library substance, which amounts to a total of 257 ms for the full substance library containing PETN, TNT, RDX and AN. The following AMSD target detection algorithm uses this result for the actual detection step, which was measured to require 82 ms per time frame and substance (329 ms for the full library). This amounts to a total processing time of 586 ms for the combination of ATGP and AMSD for the current library containing four substances.

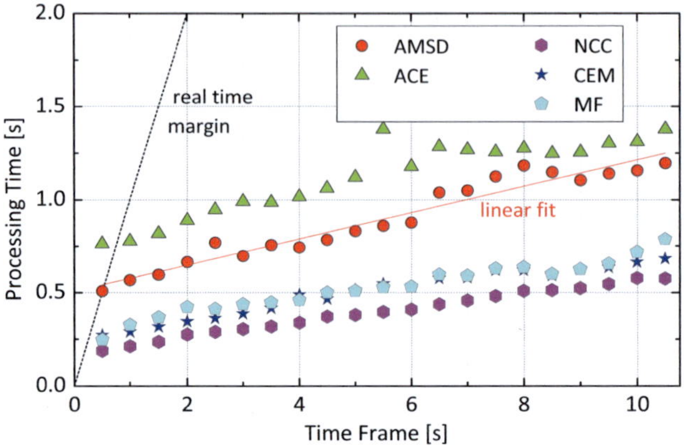

Figure 8.12: Analysis of real-time capability of the considered target detection algorithms using sequential programming. All detection algorithms require a processing time that is below the data acquisition time from 1 s and upwards. The full pixel detectors are in general faster, compared to the sub pixel detectors, among which the AMSD slightly outperforms the ACE.

As the resulting processing times are considerably shorter than the considered data frame time, we conclude that the suggested algorithms are well applicable for real-time applications in the considered scale. Furthermore, the detection process can be easily parallelized (substance detection for each substance in the library is independent to the others), hence parallel processing can be employed, to reduce the computation time by the number of available cores. Alternatively, the data frame time can be adjusted, to match the analysis time. Figure 8.12 shows the full library data processing time for various choices of data frame time and several detection algorithms, acquired on a 22 s data stream of hyperspectral data.

As expected, compared to the model based sub pixel detection algorithms Adaptive Coherence / Cosine Estimator (ACE) and AMSD the full pixel detectors require considerably less processing time. In terms of processing time, among the structured detection algorithms, the AMSD slightly outperforms the ACE. The processing time of all detection algorithms show an approximately linear growth over data frame width (i. e. observation size) with a proportionality coefficient well beyond one. We conclude that real-time operation for arbitrary library sizes can be obtained by appropriate choice of data frame width. Assuming sequential processing, all considered target detection algorithms prove to be capable of real-time operation under the given operation conditions, if a data frame width of 1 s or greater is chosen. We note that on average the Normalized Cross Correlation (NCC) target detector is at least a factor of ≈ 1.8 times faster compared to the AMSD algorithm, especially since full pixel detectors operate on single spectra, whereas sub pixel detectors require a data frame comprising multiple spectra. Hence, the NCC – or in general any member of the family of full pixel detection algorithms – might be a favorable choice in applications where no or only weak background spectral contribution is expected.

8.3.2 Conclusion

In this chapter we presented a real-time backscattering spectroscopy principle based upon a rapid wavelength-scanning MOEMS EC-QCL. The obtained measurement results indicate that the measured spectra match well to the library spectra measured by imaging backscattering spectroscopy. Furthermore we showed that the considered hyperspectral image analysis algorithms are in general well transferable to fast identification of surface contamination using the real-time spectroscopy setup. Finally, an analysis of the expected computation time showed that real-time analysis of the acquired spectroscopic data is achievable using both full pixel and sub pixel target detection algorithms.

9 Summary and Outlook

In this work, tunable infrared laser backscattering spectroscopy based on External Cavity Quantum Cascade Lasers (EC-QCLs) was presented. Broadly tunable long-wave infrared (LWIR) Quantum Cascade Lasers (QCLs) are used for wavelength selective illumination of the scene to be analyzed. The considered wavelength range from $1000\,cm^{-1}$ to $1300\,cm^{-1}$ comprises a part of the molecular fingerprint range of many organic compounds. Whereas the measurement principle is for this reason in principle applicable to detection of many other chemical substances, focus was put on detection of explosive substances in this work. The examined substances include the explosives PETN, RDX, TNT and AN.
The proposed technique of active laser spectroscopic imaging employs tunable EC-QCLs for wavelength selective illumination of the scene. The diffusely backscattered light is collected by a high performance Mercury Cadmium Telluride (MCT) LWIR image sensor. Synchronous wavelength tuning and imaging yields a hyperspectral image, where every pixel vector comprises a backscattering spectrum of a specific location in the scene.
An optimized data acquisition for the target application by identification and elimination of several systematic noise sources was developed. This includes a homogenization step that serves to correct for spatial dependent multiplicative oscillations caused by thin-film interference effects in the camera sensor in combination with the coherent illumination source. It was shown, that the spectra obtained by backscattering spectroscopy are well comparable to reflectance spectra acquired using Fourier Transform Infrared (FTIR) spectroscopy.

A comparison of FTIR microscopy spectra of single substance particles showed that the observed intra-substance spectral variability of the dominating spectral feature characteristics are insignificant, which suggests that well reproducible measurement results can be expected from macroscopic measurement techniques. An analysis of backscattering spectroscopy results proved that the Linear Mixture Model (LMM) describes the observed data well, which indicates that the measured spectra comprise a linear superposition of independent sources. The latter correspond to the spectrally distinct materials in the scene, and are referred to as *endmembers*.

Various data analysis techniques were considered for target detection in the acquired hyperspectral images. Many of these methods are based upon the LMM that – in it's *structured* form – assumes the observation data to be a superposition of the contributing endmembers and the potential target spectrum. It was therefore considered necessary, to develop a method capable of estimating the number of contributing pure background substances from the data set under test – this is referred to as *model order estimation*.

Based on the work of Wax and Kailath the Noise Adjusted Minimum Description Length (NA-MDL) model order estimation method was developed, which is applicable to any real valued multiband observation that comprise a linear superposition of independent sources. The method was analyzed in the context of the target operation conditions and compared to a selection of other model order estimation algorithms (Principal Components Analysis (PCA), Noise Subspace Projection (NSP), Second Moment Linear Dimension (SML)). On artificial data the presented NA-MDL method proved to outperform the competing approaches under a variety of conditions, including low Signal to Noise Ratio (SNR) or low spectral resolution. For the case of real-world hyperspectral image data, both NA-MDL and NSP showed to significantly overestimate the required model order estimate. Here, SML estimates proved to be closer to the expected outcome. Supported by a comparison of the covariance matrix eigenvalue distributions of artificial and real-world measurement data, this observation was accounted to correlated noise in the real-world hyperspectral images. Given the coherent nature of the illumination source, the noise can most likely be attributed to remaining speckle that was not fully eliminated by the speckle reduction optics. It is therefore anticipated that further improvement of speckle reduction techniques that are applicable in the LWIR range covered by the suggested spectroscopy methods would be beneficial for results obtained from the NA-MDL algorithm.

Based on the Adaptive Target Generation Process (ATGP), a background estimation algorithm called Adaptive Background Generation Process (ABGP) was developed that proved to be robust against model overestimation artifacts. A reliable background endmember representation is generated, even if the required model order is significantly exceeded. Noise reduction is achieved by introducing a clustering step into the procedure and utilizing the cluster means, rather than single spectra as endmembers. The background estimation procedure was analyzed using both artificial and real-world hyperspectral image data.

The discriminative power of structured target detection methods is strongly dependent upon the quality of the background endmember representation. Using artificial hyperspectral image data, a comparison of the considered target detection methods showed that the Adaptive Matched Subspace Detector (AMSD) – as widely used member of the family of structured detectors – significantly outperforms the full-pixel detection algorithms, if the AMSD background estimation method is employed. Among the unstructured sub pixel target detection methods, the Adaptive Coherence / Cosine Estimator (ACE) yielded comparatively high detection performance. These results showed to be well transferable to real-world hyperspectral images. Several examples of explosive residue detection on different substrates were provided using PETN, RDX, TNT, AN as demonstration substances. Successful detection was shown both for hyperspectral images acquired with the short-range imaging system over a distance of 1.4 m, as well as with the long-range image sensor at distances up to 17 m.

An extension to a real time capable variant of the measurement principle was presented that comprises a MOEMS External Cavity Quantum Cascade Laser (MOEMS EC-QCL) as rapid wavelength scanning illumination source. An analysis of the fast data acquisition setup showed that the expected spectral resolution should be sufficient for spectroscopy of solids with typical spectral band widths of $1\,cm^{-1}$. The data analysis algorithms showed to be capable of the substance detection on hyperspectral data acquired using this backscattering spectroscopy variant. An analysis of the required computation time proved that all considered data analysis algorithms are capable of real time detection under the given operation conditions.
At this time, the data acquisition scheme of the imaging sensor, leaves room for further optimization. During the camera sensor's integration time (max. 100 µs), the laser is operated in pulsed operation mode at a duty cycle of 17 %, which comprises the laser's duty cycle limit. Advances in continuous wave (cw) QCL chips, are expected to enable illumination of the target throughout the full integration time period. Alternatively, gated viewing can be applied to match the sensor's integration time within a frame, to the laser's illumination time. The integration time reduction serves to limit the exposure to the interfering radiation caused by the thermal background of the scene. In addition, this approach is favored to the aforementioned cw QCL solution, as in general both higher peak power and a broader tuning range can be achieved in pulsed mode compared to cw operation. Furthermore, future QCLs are expected to improve, both in terms of emission power and spectral emission range. Whereas the former is beneficial in terms of achievable SNR or can serve to increase the illuminated area, the latter is expected to further increase the detection performance and reduce potential cross-sensitivity.
The coherent nature of the laser illumination source comprises problems both in data acquisition and analysis. It is therefore desirable, to improve the speckle reduction approach, which could be achieved by e. g. employing

multiple independent illumination sources that are spatially distributed sufficiently far from each other.
The MOEMS EC-QCL based real time backscattering spectroscopy technique is to this point demonstrated using a single element detector. The data acquisition density could be however significantly improved by employing a line focus in combination with a detector array. This induces however higher complexity in signal acquisition and issues additional time constraints on the data analysis process. Furthermore, a mobile and compact spectroscopy device based on this technique, would be highly suitable for fast substance analysis in the field. This raises the demand for sensitive and robust, high performance hyperspectral data analysis algorithms that allow real-time operation within the limited hardware resources provided by embedded systems.

A Appendix

Log-Likelihood Function

We begin with the parametric covariance matrix model

$$\mathrm{R}^{(k)} = \sum_{i=1}^{k} (\lambda_i - \sigma^2)\vec{v}_i\vec{v}_i^\top + \sigma^2\mathrm{I} \tag{A.1}$$

and recall the definition of the parameter vector:

$$\vec{\theta}^{(k)} = (\lambda_1, \ldots, \lambda_k, \sigma^2, \vec{v}_1^\top, \ldots, \vec{v}_k^\top)^\top. \tag{A.2}$$

Given an observation $\mathrm{X} = \{\vec{\mathrm{x}}_i\}_{i=1}^N$ and assuming a multivariate Normal distribution we are searching for an explicit form of the log-likelihood function $\mathscr{L}(\mathrm{X} \mid \vec{\theta}^{(k)})$.

$$\begin{aligned}
\mathscr{L}(\vec{\mathrm{x}}_1, \ldots, \vec{\mathrm{x}}_N \mid \vec{\theta}^{(k)}) &= \log\left[p(\vec{\mathrm{x}}_1, \ldots, \vec{\mathrm{x}}_N \mid \vec{\theta}^{(k)})\right] = \sum_{i=1}^{N} \log\left[p(\vec{\mathrm{x}}_i \mid \vec{\theta}^{(k)})\right] \\
&= \sum_{i=1}^{N} \log\left[(2\pi)^{-\frac{p}{2}} \cdot \det \mathrm{R}^{(k)^{-\frac{1}{2}}} \cdot \exp\left[-\frac{1}{2}(\vec{\mathrm{x}}_i - \vec{\mu})^\top \mathrm{R}^{(k)^{-1}} (\vec{\mathrm{x}}_i - \vec{\mu})\right]\right] \\
&= -\frac{1}{2} \cdot \left[N \log(2\pi)^p + N \log \det \mathrm{R}^{(k)} + \sum_{i=1}^{N} (\vec{\mathrm{x}}_i - \vec{\mu})^\top \mathrm{R}^{(k)^{-1}} (\vec{\mathrm{x}}_i - \vec{\mu})\right] \\
&= -\frac{1}{2} \cdot \left[N \log(2\pi)^p + N \log \det \mathrm{R}^{(k)} + \sum_{i=1}^{N} \mathrm{tr}\left[(\vec{\mathrm{x}}_i - \vec{\mu})^\top \mathrm{R}^{(k)^{-1}} (\vec{\mathrm{x}}_i - \vec{\mu})\right]\right] \\
&= -\frac{1}{2} \cdot \left[N \log(2\pi)^p + N \log \det \mathrm{R}^{(k)} + \mathrm{tr}\left[\mathrm{R}^{(k)^{-1}} \cdot \sum_{i=1}^{N} (\vec{\mathrm{x}}_i - \vec{\mu})(\vec{\mathrm{x}}_i - \vec{\mu})^\top\right]\right] \\
&= -\frac{N}{2}\left[\log(2\pi)^p + \log \det \mathrm{R}^{(k)} + \mathrm{tr}\left[\mathrm{R}^{(k)^{-1}} \cdot \hat{\mathrm{R}}\right]\right],
\end{aligned} \tag{A.3}$$

where once more $\hat{\mathrm{R}}$ denotes the sample covariance matrix. We use equation A.1 to calculate the logarithmic determinant of $\mathrm{R}^{(k)}$ and obtain:

$$\log\det\mathrm{R}^{(k)} = \log\left[\prod_{i=1}^{p}\lambda_i\right] = \sum_{i=1}^{k}\log\lambda_i + (p-k)\cdot\log\sigma^2 \tag{A.4}$$

The trace term can be evaluated to:

$$\mathrm{tr}\left[{\mathrm{R}^{(k)}}^{-1}\cdot\hat{\mathrm{R}}\right] = \sum_{i=1}^{k}\frac{l_i}{\lambda_i} + \sum_{i=k+1}^{p}\frac{l_i}{\sigma^2} \tag{A.5}$$

Putting these results together, we get the closed-form expression:

$$\begin{aligned}\mathscr{L}(\vec{\mathrm{x}}_1,\ldots,\vec{\mathrm{x}}_N\,|\,\vec{\theta}^{(k)}) &= -\frac{N}{2}\left[\log(2\pi)^p + \log\det\mathrm{R}^{(k)} + \mathrm{tr}\left[{\mathrm{R}^{(k)}}^{-1}\cdot\hat{\mathrm{R}}\right]\right] \\ &= -\frac{N}{2}\left[\log(2\pi)^p + \sum_{i=1}^{k}\left[\log\lambda_i + \frac{l_i}{\lambda_i}\right] + (p-k)\cdot\log\sigma^2 + \sum_{i=k+1}^{p}\frac{l_i}{\sigma^2}\right].\end{aligned} \tag{A.6}$$

The Maximum Likelihood Estimates (MLEs) that maximize the log-likelihood function are [2]

$$\begin{aligned}\hat{\lambda}_i &= l_i,\, i=1,\ldots,k,\, l_i\in\mathbb{R} \\ \hat{\sigma}^2 &= \frac{1}{p-k}\sum_{i=k+1}^{p} l_i \\ \hat{\vec{\mathrm{v}}}_i &= \vec{\mathrm{c}}_i,\, i=1,\ldots,k,\, \vec{\mathrm{c}}_i\in\mathbb{R}^p.\end{aligned} \tag{A.7}$$

It is left, to substitute these in equation A.6:

$$\begin{aligned}&\mathscr{L}(\vec{\mathrm{x}}_1,\ldots,\vec{\mathrm{x}}_N\,|\,\vec{\theta}^{(k)}) \\ &= -\frac{N}{2}\left[\log(2\pi)^p + \sum_{i=1}^{k}[\log\lambda_i + 1] + (p-k)\cdot\log\sum_{i=k+1}^{p}\frac{l_i}{p-k} + \frac{1}{\hat{\sigma}^2}\sum_{i=k+1}^{p} l_i\right] \\ &= -\frac{N}{2}\left[\log(2\pi)^p + k + \sum_{i=1}^{k}\log\lambda_i + (p-k)\cdot\log\sum_{i=k+1}^{p}\frac{l_i}{p-k} + (p-k)\right]\end{aligned}$$

$$= -\frac{N}{2}\left[p + \log(2\pi)^p + \sum_{i=1}^{k}\log\lambda_i + (p-k)\cdot\log\sum_{i=k+1}^{p}\frac{l_i}{p-k}\right]$$

$$= -\frac{N}{2}\left[\sum_{i=1}^{k}\log\lambda_i + \log\left[\sum_{i=k+1}^{p}\frac{l_i}{p-k}\right]^{(p-k)}\right] + c. \tag{A.8}$$

List of Figures

List of Tables

Substance List

AN	Ammoniumnitrate
ANFO	Ammonium Nitrate / Fuel Oil
DNT	Dinitrotoluene
PETN	Pentaerythritol tetranitrate
RDX	Research Department Formula X
TATP	Tri-cyclic Acetone Peroxide
TNT	Trinitrotoluene

Physical Acronyms

ATR	Attenuated Total Reflectance
EC-QCL	External Cavity Quantum Cascade Laser
FTIR	Fourier Transform Infrared
InAlAs	Indium Aluminum Arsenide
InGaAs	Indium Gallium Arsenide
InP	Indium Phosphide
KTP	Potassium Titanyl Phosphate
LWIR	long-wave infrared
MCT	Mercury Cadmium Telluride
MIR	mid-infrared
MOEMS	Micro Opto Electromechanical System
MOEMS EC-QCL	MOEMS External Cavity Quantum Cascade Laser
OPO	Optical Parametric Oscillator
QCL	Quantum Cascade Laser
ZnSe	Zink Selenide

Mathematical Acronyms

ABGP	Adaptive Background Generation Process
ACE	Adaptive Coherence / Cosine Estimator
AMSD	Adaptive Matched Subspace Detector
AMSD-ABGP	AMSD detector with ABGP background extraction
ATGP	Adaptive Target Generation Process
AWGN	Additive White Gaussian Noise
CEM	Constrained Energy Minimization
CFAR	Constant False Alarm Rate
FAR	False Alarm Rate
HFC	Harsanyi-Farrand-Chang
LMM	Linear Mixture Model
MDL	Minimum Description Length
MF	Matched Filter
MLE	Maximum Likelihood Estimate
NA-MDL	Noise Adjusted Minimum Description Length
NCC	Normalized Cross Correlation
NSP	Noise Subspace Projection

OSP	Orthogonal Subspace Projection
PCA	Principal Components Analysis
ROC	Receiver Operating Characteristic
SML	Second Moment Linear Dimension
SNR	Signal to Noise Ratio
TPR	True Positive Rate

Bibliography

[1] Adams, R. W., Vijayraghavan, K., Wang, Q. J., Fan, J., Capasso, F., Khanna, S. P., Davies, a. G., Linfield, E. H., and Belkin, M. a. (2010). GaAs/$Al_{15}Ga_{85}$As terahertz quantum cascade lasers with double-phonon resonant depopulation operating up to 172 K. *Appl. Phys. Lett.*, 97(13):131111.

[2] Anderson, T. W. (1963). Asymptotic theory for principal component analysis. *Ann. Math. Stat.*, pages 122–148.

[3] Bajorski, P. (2009). Does virtual dimensionality work in hyperspectral images? In *Proc. SPIE*, volume 7334, pages 73341J—-73341J.

[4] Bajorski, P. (2011). Second moment linear dimensionality as an alternative to virtual dimensionality. *IEEE Trans. Geosci. Remote Sens.*, 49(2):672–678.

[5] Barron, A., Rissanen, J., and Yu, B. (1998). The minimum description length principle in coding and modeling. *IEEE Trans. Inf. Theory*, 44(6):2743–2760.

[6] Beyer, T., Braun, M., and Lambrecht, A. (2003). Fast gas spectroscopy using pulsed quantum cascade lasers. *J. Appl. Phys.*, 93(6):3158–3160.

[7] Broadwater, J. (2008). Effects of endmember dimensionality on subpixel detection performance. In *Int. Geosci. Remote Sens. Symp.*, volume 2, pages II—-613.

[8] Capasso, F. (2010). High-performance midinfrared quantum cascade lasers. *Opt. Eng.*, 49(11):111102.

[9] Chang, C. I. and Du, Q. (1999). Interference and noise-adjusted principal components analysis. *IEEE Trans. Geosci. Remote Sens.*, 37(5):2387–2396.

[10] Chang, C. I. and Du, Q. (2004). Estimation of number of spectrally distinct signal sources in hyperspectral imagery. *IEEE Trans. Geosci. Remote Sens.*, 42(3):608–619.

[11] Chang, C. I., Wu, C. C., Liu, W., and Ouyang, Y. C. (2006). A new growing method for simplex-based endmember extraction algorithm. *IEEE Trans. Geosci. Remote Sens.*, 44(10):2804–2819.

[12] Chang, C.-I. and Xiong, W. (2010). High-order statistics Harsanyi-Farrand-Chang method for estimation of virtual dimensionality. In *Proc. SPIE*, volume 7810, page 78100D.

[13] Colthup, N. B., Daly, L. H., and Wiberley, S. E. (1975). *Introduction to Infrared and Raman Spectroscopy*. Academic Press, Inc., London.

[14] Faist, K., Hofstetter, D., Beck, M., Aellen, T., Rochat, M., and Blaser, S. (2002). Bound-to-continuum and two-phonon resonance, quantum-cascade lasers for high duty cycle, high-temperature operation. *IEEE J. Quantum Electron.*, 38(6):533–546.

[15] Fano, R. (1949). *The transmission of ilnformation*. Massachusetts Institute of Technology, Cambridge.

[16] Farrand, W. H. and Harsanyi, J. C. (1997). Mapping the Distribution of Mine Tailings in the Coeur d'Alene River Valley, Idaho, through the Use of a Constrained Energy Minimization Technique. *Remote Sens. Environ.*, 59(1):64–67.

[17] Fuchs, F., Hugger, S., Kinzer, M., Hinkov, B., Aidam, R., Bronner, W., Lösch, R., Yang, Q., Degreif, K., Schnürer, F., and Others (2010). Imaging stand-off detection of explosives by quantum cascade laser based backscattering spectroscopy. In *Proc. SPIE*, volume 7808, page 780810.

[18] Furstenberg, R., Grosser, J., Kendziora, C. A., Papantonakis, M. R., Nguyen, V., and McGill, R. A. (2011). Modeling of laser-analyte-substrate interaction in photo-thermal infrared imaging and laser trace vaporization. In *Proc. SPIE*, volume 8013, page 801318.

[19] Furstenberg, R., Kendziora, C. A., Papantonakis, M. R., Nguyen, V., and McGill, R. A. (2014). The challenge of changing signatures in infrared stand-off detection of trace explosives. In *Proc. SPIE*, volume 9073, page 90730M.

[20] Furstenberg, R., Kendziora, C. A., Stepnowski, J., Stepnowski, S. V., Rake, M., Papantonakis, M. R., Nguyen, V., Hubler, G. K., and McGill, R. A. (2008). Stand-off detection of trace explosives via resonant infrared photothermal imaging. *Appl. Phys. Lett.*, 93(22):224103.

[21] Gares, K. L., Hufziger, K. T., Bykov, S. V., and Asher, S. A. (2016). Review of explosive detection methodologies and the emergence of standoff deep UV resonance Raman. *J. Raman Spectrosc.*, 47(1):124–141.

[22] Goodman, J. W. (2007). *Speckle Phenomena in Optics: Theory and Applications*. Roberts and Company Publishers.

[23] Gottfried, J. L., De Lucia, F. C., Munson, C. A., and Miziolek, A. W. (2009). Laser-induced breakdown spectroscopy for detection of explosives residues: a review of recent advances, challenges, and future prospects. *Anal. Bioanal. Chem.*, 395(2):283–300.

[24] Grahmann, J., Merten, A., Ostendorf, R., Fontenot, M., Bleh, D., Schenk, H., and Wagner, J. (2014). Tunable External Cavity Quantum Cascade Lasers (EC-QCL): an application field for MOEMS based scanning gratings. In *Proc. SPIE*, volume 8977, page 897708.

[25] Grünwald, P. (2005). *A tutorial introduction to the minimum description length principle*. MIT Press.

[26] Grünwald, P. (2007). *The Minimum Description Length Principle*. MIT Press.

[27] Harsanyi, J., Farrand, W., and Chang, C.-I. (1992). Determining the number and identity of spectral endmembers: An integrated approach using Neyman-Pearson eigenthresholding and iterative constrained RMS error minimization. In *Proc. Themat. Conf. Geol. Remote Sens.*, volume 1, pages 395–408.

[28] Harsanyi, J. C. and Chang, C. I. (1994). Hyperspectral image classification and dimensionality reduction: An orthogonal subspace projection approach. *IEEE Trans. Geosci. Remote Sens.*, 32(4):779–785.

[29] Heinz, D. C. and Chang, C. I. (2001). Fully constrained least squares linear spectral mixture analysis method for material quantification in hyperspectral imagery. *IEEE Trans. Geosci. Remote Sens.*, 39(3):529–545.

[30] Jarvis, J., Fuchs, F., Hugger, S., Ostendorf, R., Butschek, L., Yang, Q., Dreyhaupt, A., Grahmann, J., and Wagner, J. (2016). Hyperspectral image analysis for standoff trace detection using IR laser spectroscopy. In *Proc. SPIE*, volume 9824, page 98240V.

[31] Jarvis, J., Haertelt, M., Hugger, S., Butschek, L., Fuchs, F., Ostendorf, R., Wagner, J., and Beyerer, J. (2017). Hyperspectral Data Acquisition

and Analysis in Imaging and Real-Time Active MIR Backscattering Spectroscopy. *Adv. Opt. Technol.*, 6(2):85–93.

[32] Juds, S. (1988). *Photoelectric Sensors and Controls: Selection and Application, First Edition*, volume 63. CRC Press.

[33] Kay, S. (1998). *Fundamentals of Statistical Signal Processing: Detection theory*. Prentice Hall, New Jersey, vol. 2 edition.

[34] Kendziora, C. A., Furstenberg, R., Papantonakis, M., Nguyen, V., Borchert, J., Byers, J., and McGill, R. A. (2013). Infrared photothermal imaging of trace explosives on relevant substrates. In *Proc. SPIE*, volume 8373, pages 83732H–1.

[35] Kendziora, C. A., Furstenberg, R., Papantonakis, M., Nguyen, V., Byers, J., and Andrew McGill, R. (2015). Infrared photothermal imaging spectroscopy for detection of trace explosives on surfaces. In *Appl. Opt.*, volume 54, page F129.

[36] Kendziora, C. a., Furstenberg, R., Papantonakis, M., Nguyen, V., Stepnowski, J., and McGill, R. A. (2010). Advances in stand-off detection of trace explosives by infrared photo-thermal imaging. In *Proc. SPIE*, volume 7664, page 76641J.

[37] Kim, S., Lee, D., Liu, X., Van Neste, C. W., and Thundat, T. (2014). Point and standoff detection of trace explosives using quantum cascade lasers. In *Proc. SPIE*, volume 9083, page 90832N.

[38] Laurenzis, M., Lutz, Y., Christnacher, F., Matwyschuk, A., and Poyet, J.-M. (2012). Homogeneous and speckle-free laser illumination for range-gated imaging and active polarimetry. *Opt. Eng.*, 51(6):061302.

[39] Lee, B. G., Zhang, H. A., Pfluegl, C., Diehl, L., Capasso, F., Belkin, M. A., Fischer, M., Wittmann, A., and Faist, J. (2009). Broadband

distributed-feedback quantum cascade laser array operating from 8.0 to 9.8 μm. *IEEE Photonics Technol. Lett.*, 21(13):914–916.

[40] Liu, X., Van Neste, C. W., Gupta, M., Tsui, Y. Y., Kim, S., and Thundat, T. (2014). Standoff reflection–absorption spectra of surface adsorbed explosives measured with pulsed quantum cascade lasers. *Sensors Actuators B Chem.*, 191:450–456.

[41] Manolakis, D. (2003). Detection algorithms for hyperspectral imaging applications: A signal processing perspective. In *IEEE Work. Adv. Tech. Anal. Remote. Sensed Data*, pages 378–384. IEEE.

[42] Manolakis, D., Shaw, G., and Keshava, N. (2000). Comparative analysis of hyperspectral adaptive matched filter detectors. In *SPIE Proc. Ser.*, pages 2–17.

[43] Manolakis, D., Siracusa, C., and Shaw, G. (2001). Hyperspectral subpixel target detection using the linear mixing model. *IEEE Trans. Geosci. Remote Sens.*, 39(7):1392–1409.

[44] Maulini, R., Mohan, A., Giovannini, M., Faist, J., and Gini, E. (2006). External cavity quantum-cascade laser tunable from 8.2 to 10.4$\sim\mu$m using a gain element with a heterogeneous cascade. *Appl. Phys. Lett.*, 88(May 2007):201113.

[45] Nordberg, M., Åkeson, M., Östmark, H., and Carlsson, T. E. (2011). Stand-off detection of explosive particles by imaging Raman spectroscopy. In *Proc. SPIE*, volume 8017, page 80171B.

[46] Nordberg, M., Ceco, E., Wallin, S., and Östmark, H. (2012). Detection limit of imaging Raman spectroscopy. In *Proc. SPIE*, volume 8357, page 83571H. International Society for Optics and Photonics.

[47] Ostendorf, R., Butschek, L., Hugger, S., Fuchs, F., Yang, Q., Jarvis, J., Schilling, C., Rattunde, M., Merten, A., Grahmann, J., Boskovic, D.,

Tybussek, T., Rieblinger, K., and Wagner, J. (2016). Recent Advances and Applications of External Cavity-QCLs towards Hyperspectral Imaging for Standoff Detection and Real-Time Spectroscopic Sensing of Chemicals. *Photonics*, 3(2):28.

[48] Östmark, H., Nordberg, M., and Carlsson, T. E. (2011). Stand-off detection of explosives particles by multispectral imaging Raman spectroscopy. *Appl. Opt.*, 50(28):5592.

[49] Pettersson, A., Johansson, I., Wallin, S., Nordberg, M., and Östmark, H. (2009). Near Real-Time Standoff Detection of Explosives in a Realistic Outdoor Environment at 55 m Distance. *Propellants, Explos. Pyrotech.*, 34(4):297–306.

[50] Rauter, P. and Capasso, F. (2015). Multi-wavelength quantum cascade laser arrays. *Laser & Photonics Rev.*, 9(5):452–477.

[51] Ren, H. and Chang, C. I. (2003). Automatic spectral target recognition in hyperspectral imagery. *IEEE Trans. Aerosp. Electron. Syst.*, 39(4):1232–1249.

[52] Rissanen, J. (1978). Modeling by shortest data description. *Automatica*, 14(5):465–471.

[53] Roger, R. and Arnold, J. (1996). Reliably estimating the noise in AVIRIS hyperspectral images. *Int. J. Remote Sens.*, 17(10):1951–1962.

[54] Scharf, L. L. and Friedlander, B. (1994). Matched subspace detectors. *Signal Process. IEEE Trans.*, 42(8):2146–2157.

[55] Schwarz, G. (1978). Estimating the Dimension of a Model. *Ann. Stat.*, 6(2):461–464.

[56] Shannon, C. E. (1948). A mathematical theory of communication. *Bell Syst. Tech. J.*, 27(623-656):379–423.

[57] Smith, M. O., Adams, J. B., and Johnson, P. E. (1985). Quantitative determination of mineral types and abundances from reflectance spectra using principal components analysis. *J. Geophys. Res.*, 90(S02):C797–C804.

[58] Thai, B. and Healey, G. (2002). Invariant subpixel material detection in hyperspectral imagery. *IEEE Trans. Geosci. Remote Sens.*, 40(3):599–608.

[59] Torrent, J. and Barrón, V. (2008). Diffuse reflectance spectroscopy. In *Methods Soil Anal.*, number 5, chapter Diffuse Re, pages 367–385.

[60] Van Neste, C. W., Liu, X., Gupta, M., Kim, S., Tsui, Y., and Thundat, T. (2012). Standoff detection of explosive residues on unknown surfaces. In *Proc. SPIE*, volume 8373, page 83732F.

[61] Van Neste, C. W., Senesac, L. R., and Thundat, T. (2008). Standoff photoacoustic spectroscopy. *Appl. Phys. Lett.*, 92(23):234102.

[62] Walther, C., Fischer, M., Scalari, G., Terazzi, R., Hoyler, N., and Faist, J. (2007). Quantum cascade lasers operating from 1.2 to 1.6 THz. *Appl. Phys. Lett.*, 91(13):131122.

[63] Wax, M. and Kailath, T. (1985). Detection of signals by information theoretic criteria. *IEEE Trans. Acoust. Speech Signal Process.*, 33(2):387–392.

[64] Wysocki, G., Lewicki, R., Curl, R. F., Tittel, F. K., Diehl, L., Capasso, F., Troccoli, M., Hofler, G., Bour, D., Corzine, S., Maulini, R., Giovannini, M., and Faist, J. (2008). Widely tunable mode-hop free external cavity quantum cascade lasers for high resolution spectroscopy and chemical sensing. In *Appl. Phys. B Lasers Opt.*, volume 92, pages 305–311.

[65] Yamamoto, K. and Ishida, H. (1994). Optical theory applied to infrared spectroscopy. *Vib. Spectrosc.*, 8(1):1–36.

SCHRIFTENREIHE AUTOMATISCHE SICHTPRÜFUNG UND BILDVERARBEITUNG (ISSN 1866-5934)

Herausgeber: Prof. Dr.-Ing. Jürgen Beyerer

Band 1 **JONATHAN BALZER**
Regularisierung des Deflektometrieproblems Grundlagen und Anwendung. 2008
ISBN 978-3-86644-230-6

Band 2 **IOANA GHEŢA**
Fusion multivariater Bildserien am Beispiel eines Kamera-Arrays. 2011
ISBN 978-3-86644-684-7

Band 3 **STEFAN BRUNO WERLING**
Deflektometrie zur automatischen Sichtprüfung und Rekonstruktion spiegelnder Oberflächen. 2011
ISBN 978-3-86644-687-8

Band 4 **JAN WASSENBERG**
Efficient Algorithms for Large-Scale Image Analysis. 2012
ISBN 978-3-86644-786-8

Band 5 **MARTIN GRAFMÜLLER**
Verfahrensfortschritte in der robusten Echtzeiterkennung von Schriftzeichen. 2013
ISBN 978-3-86644-979-4

Band 6 **JÜRGEN BRAUER**
Human Pose Estimation with Implicit Shape Models. 2014
ISBN 978-3-7315-0184-8

Band 7 **MARKUS MÜLLER**
Szeneninterpretation unter Verwendung multimodaler Sensorik und Salienzmaßen. 2017
ISBN 978-3-7315-0240-1

Band 8 **ROBIN GRUNA**
Beleuchtungsverfahren zur problemspezifischen Bildgewinnung für die automatische Sichtprüfung. 2015
ISBN 978-3-7315-0313-2

Band 9 **THOMAS STEPHAN**
Beitrag zur Unterwasserbildrestauration. 2017
ISBN 978-3-7315-0579-2

Die Bände sind unter www.ksp.kit.edu als PDF frei verfügbar oder als Druckausgabe bestellbar.

Band 10 **JAN-PHILIP JARVIS**
A Contribution to Active Infrared Laser Spectroscopy for Remote Substance Detection. 2017
ISBN 978-3-7315-0725-3

Die Bände sind unter www.ksp.kit.edu als PDF frei verfügbar oder als Druckausgabe bestellbar.